SpringerBriefs in Electrical and Computer Engineering

Howard H. Yang · Tony Q.S. Quek

Massive MIMO Meets Small Cell

Backhaul and Cooperation

Howard H. Yang
Singapore University of Technology
and Design
Singapore
Singapore

Tony Q.S. Quek
Singapore University of Technology
and Design
Singapore
Singapore

ISSN 2191-8112 ISSN 2191-8120 (electronic)
SpringerBriefs in Electrical and Computer Engineering
ISBN 978-3-319-43713-2 ISBN 978-3-319-43715-6 (eBook)
DOI 10.1007/978-3-319-43715-6

Library of Congress Control Number: 2016946961

Printed on acid-free paper

This Springer imprint is published by Springer Nature
The registered company is Springer International Publishing AG Switzerland

Preface

The current growth rate of wireless data exceeds both spectral efficiency advances and availability of new wireless spectrum; a trend towards network densification is essential to respond adequately to the continued surge in mobile data traffic. To this end, there are two common ideas to densify the network, one is by aggregating lots of antennas at the base station to achieve large diversity gain, termed massive multiple-input-multiple-output (MIMO), and the other one is by spreading antennas into the network and form small autonomous regions to provide better path loss, known as small cell network.

The focus of this book is on combining these two techniques and to investigate a better utilization of the excessive spatial dimensions to improve network performance. Particularly, we point out two directions that the large number of antennas can be used for: (1) interference suppression, where we propose a linear precoding scheme termed cell-edge aware zero forcing (CEA-ZF) that exploits the extra degrees of freedom from the large base station antenna array to mitigate inter-cell interference at cell-edge neighboring users; (2) wireless backhaul, where we propose using the massive antenna array at macro base stations to simultaneously serve several small access points within their coverage by spatial multiplexing, thus connecting different tiers in a small cell network via wireless backhaul and perform an energy-efficient design. In order to quantify the performance of our proposed schemes, we combine random matrix theory and stochastic geometry to develop an analytical framework that accounts for all the key features of a network, including number of antenna array, base station density, inter-cell interference, random base station deployment, and network traffic load. The analysis enables us to explore the impact from different network parameters through numerical analysis. Our results show that on the one hand, CEA-ZF outperforms conventional zero forcing in terms of coverage probability, aggregated per cell rate, and edge user rate, demonstrating it as a more effective precoding scheme to achieve better coverage probability in massive MIMO cellular networks. On the other hand, we show that a two-tier small cell network with wireless backhaul can be significantly more energy-efficient than a one-tier cellular network. However, this requires the bandwidth division between radio access links and wireless backhaul to be optimally designed according to the load conditions.

We would like to thank Dr. Giovanni Geraci from Bell Labs, Ireland, and Prof. Jeffrey G. Andrews from University of Texas at Austin, for their comments in improving the quality of this book.

Singapore   Howard H. Yang
2016   Tony Q.S. Quek

Contents

Acronyms

BD	Block diagonalization
BS	Base station
CCDF	Complementary cumulative density function
CDF	Cumulative density function
CEA-ZF	Cell-edge aware zero forcing
CEU-Zf	Cell-edge unaware zero forcing
CSI	Channel state information
DL	Downlink
i.i.d.	Independent and identically distributed
MBS	Macro cell base station
MIMO	Multiple-input-multiple-output
MMSE	Minimum-mean-square-error
MRT	Maximum ratio transmission
Pdf	Probability density function
PPP	Poisson point process
QoS	Quality of Service
RB	Resource block
RSRP	Reference signal received power
SAP	Small cell access point
SINR	Signal-to-interference-plus-noise ratio
SIR	Signal-to-interference ratio
TDD	Time-division duplexing
UE	User equipment
UL	Uplink
ZF	Zero forcing

Chapter 1
Introduction

Abstract In this chapter, we present a general overview for two promising candidates of next generation wireless technologies, the massive multiple-input multiple-output (MIMO) system and the small cell networks. After respectively reviewing their concepts, advantages, and challenges, we provide the motivation and contribution of this book.

1.1 Background

Driven by new generation of wireless devices and the proliferation of bandwidth-intensive applications, user data traffic and the corresponding network load are increasing in an exponential manner, leading challenges in wireless industry to support higher data rates and ensure a consistent quality of service (QoS) throughout the network. To address this challenge, it requires network capacity to be increased by a factor of thousand over next ten years. Since spectral resources are scarce, there is a broad consensus to achieve this through network densification, i.e., deploying more antennas per unit area into the network.

In general, there are two approaches to densify the network [1]. The first direction is by aggregating more antennas at the existing base stations (BSs) to spatially multiplex user equipments (UEs) on the same time–frequency resource block. If the number of antennas largely exceeds the number of actively transmitting or receiving UEs per cell, it is epitomized as massive MIMO, where tremendous spatial degrees of freedom can overcome the fluctuation in wireless channel and bring a significant power gain to the system [2–4]. The second approach is termed small cells, which suggests to deploy antennas in a distributed manner and form autonomous regions with each covering a smaller area and serving much fewer UEs than a traditional macro base station [5, 6]. By shrinking the cell range, UEs not only benefit from better path loss, but also less competitors to share spectrum resource in the same cell, thus both network throughput and spatial reuse can be improved. Either approach will eventually lead the cellular network to be operating in a regime where number of serving antennas largely exceeds number of UEs. Such excessive degrees of freedom not only provide opportunities to attain higher capacity, but also power new direc-

© The Author(s) 2017   1
H.H. Yang and T.Q.S. Quek, *Massive MIMO Meets Small Cell*, SpringerBriefs
in Electrical and Computer Engineering, DOI 10.1007/978-3-319-43715-6_1

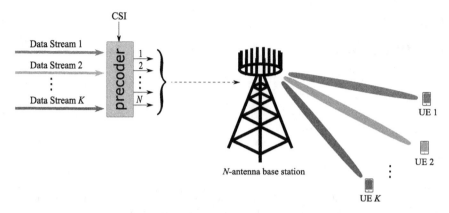

Fig. 1.1 Example of a downlink massive MIMO system, where base station antenna number N is orders larger than the number of UEs K. Data streams are simultaneously transmitted to all UEs in the cell

tions for enhanced network design to achieve better energy efficiency or improved coverage for cell-edge UEs [7, 8]. To this end, we first detail the concept of massive MIMO and small cell networks and then outline our contributions in this book that improve energy efficiency, coverage, and cell-edge user rate.

1.1.1 Massive MIMO System

Massive MIMO is a form of multi-user MIMO where BS deploys an antenna array with hundreds of active elements to serve tens of active UEs in the same time–frequency resource block [3, 9]. Figure 1.1 illustrates the concept of a typical downlink massive MIMO system, where multiple data streams for different intended UEs are precoded using channel state information (CSI) estimated from training phase, and sent out simultaneously from the BS. It has been shown that as the BS antenna array scales up, beams steered at different UEs will eventually be almost orthogonal to each other [2, 10]. In this regard, massive MIMO enables each of its UEs to enjoy a wireless channel that has high power gain and small crosstalk.

Several nice properties emerge when MIMO array are made large, which are summarized as follows:

- The vast number of antenna elements enable BS using simple linear precoding/decoding schemes, such as maximum ratio transmission (MRT) or zero forcing (ZF), to achieve the optimal channel capacity [11].
- The action of the law of large numbers smoothens out frequency dependencies in the channel, which enables an easy design of power control to improve spectral efficiency without regarding the short term fading channels [12].

- The large antenna array allows BS to concentrate its transmit power at the receivers by forming narrow beams. In this way, radiated power at BS can also be significantly reduced [13, 14].

Besides, massive MIMO also has many potential trails as reducing latency on the air interface, simplifies the multiple access layer, and increases the system robustness [3].

While very promising, massive MIMO still presents a number of research challenges. For instance, since coherence time of a wireless channel is naturally finite, there are only limited amount of orthogonal pilot sequences that can be assigned to devices for acquiring channel knowledge. Consequently, these pilot sequences have to be reused for all cells in the network, which inevitably contaminates the estimated channel, resulting in a crucial limitation to the system performance [11, 15, 16]. More importantly, though crosstalk-free channels is attainable per cell with massive MIMO, UEs located at the cell edges still surfer from interference generated by BSs transmitting in other cells, which significantly limits their experience [16–18]. In this book, by elaborating the high number of antennas at base stations, we propose a scheme that can coordinate the inter-cell interference without the need for orthogonalizing resources over time or frequency.

1.1.2 Small Cell Networks

On a separate track, modern wireless devices, such as smartphones, tablets and laptops, are generating more indoor traffic than outdoor, leading to an inhomogeneous data demand across the entire network. However, the conventional cellular networks are designed to cover large areas and optimized under homogeneous traffic profile, thus facing the challenge to meet such unbalanced traffic profile from different geographical areas [5]. To this end, there emerges a trend where more and more small cellular access points are deployed into residential homes, subways, and offices, such that satisfactory user experience is achievable. This type of network architecture, where macrocell network is overlaid with a mix of lower power cells, is commonly referred to as small cell network [5, 6, 19].

By deploying a large amount of lower power network nodes that covers a small area, traffic from macrocells can be offloaded to small access points which located in shorter distance to the end UEs. In this regard, small cell network has the advantage to not only improve the indoor coverage, but also boost the spectral efficiency per unit area via spatial reuse [6, 20]. Furthermore, the small access points can be deployed with relative low network overhead, and have high potential for bringing in an energy efficient design to future wireless networks [21–23].

A typical architecture of small cell network is illustrated in Fig. 1.2, where BSs of different types coexist in the network to serve different users. The role of different users are summarized as follows [6]:

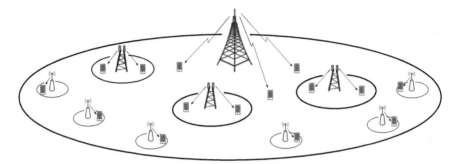

Fig. 1.2 Example of a small cell network consists of macrocell, pico cells, and femto cells. Different type of cells serve different amount of UEs

- Macrocell base stations, which are conventionally installed by operators to provide open public access. These base stations are usually destined to provide a guaranteed minimum data rate under limited delay constraint and maximum tolerable outage. They typically emit up to 46 dBm, covering a wide area on the order of few kilometers and serving thousands of customers.
- Pico cell access points, which are operator-installed low power cell towers. Pico cells are mainly deployed in places that have insufficient macro penetration, with the purpose to infill the outdoor or indoor coverage. Their transmit power generally range from 23–30 dBm, providing a coverage area around 300 m and serving a few tens of users.
- Femto cell access points, which are user-deployed access points. Femto cells are usually low cost and operating with low power, they are commonly used to offload data traffic and serve a dozen of active users in homes or enterprises. Typically, the coverage range of femto cell is less than 50 m and the transmit power is less than 23 dBm.

Small cell networks entail a shift of paradigm over the traditional cellular network, where centralized macrocells are divided into more autonomous, uncoordinated, and intelligent small cells. Though such paradigm shift provides excellent opportunities for network enhancement, several challenges also comes along. For instance, the interference alignment in co-channel deployment and hand over between different cells all act as a key limiting factor for capacity [6, 24, 25]. Besides, because of the complex topology of the various types of coexisting cells in small cell network, backhaul design poses a more challenging issue. For instance, pico cells may require to access utility infrastructure with wired backhauling, which may be potentially expensive. As for the femto cells, which in contrast use low-cost backhauling, may face difficulties to QoS since backhaul relies on users' broadband connections. Hence, operators need to carefully deploy backhaul for small cell network in a way that identifies the most cost effective and guarantees the QoS [8, 26]. Inspired by the large spatial dimensions in massive MIMO, we propose equipping macrocell base stations with large antenna arrays in a small cell network. In this sense, the tremendous diversity gain brings good opportunity for wireless backhaul in small cell network, where macro base stations have dedicated backhaul to the core network, while small access points can aggregate their traffic, and send to their closest macro base stations via wireless link. In this book, we will investigate an energy efficient design to such small cell network with wireless backhaul.

1.2 Book Outline

In the following two chapters, we mainly discuss two approaches of using the large antenna arrays, i.e., interference suppression at cell edge and wireless backhaul in small cell networks, to improve network performance.

In Chap. 2, we propose a linear precoding scheme that exploits the excessive spatial dimension to suppress downlink inter-cell interference at adjacent cell edge, in a massive MIMO network. Specifically, each BS acquires CSI of their own UEs within the cell and neighboring UEs at the cell edge, with the proposed precoding scheme, BS then sacrifices certain degrees of freedom to suppress interference toward the neighboring users and uses the remaining spatial dimensions for multiplexing gain. Through an analysis that accounts for both large antenna arrays and random BS topology, we demonstrate that if our precoding scheme can be used throughout the entire network in a distributed manner, both coverage and cell-edge rate can be significantly improved.

In Chap. 3, we investigate an energy efficient design in a small cell network, where macro base stations are equipped with large antenna array, and small access points connect to their closest macro base stations via wireless backhaul. By combining random matrix theory and stochastic geometry, we develop a general framework for the analysis, which takes a complete treatment of uplink and downlink transmissions, spatial multiplexing, and resource allocation between radio access links and backhaul. Our results show that irrespective of the deployment strategy, it is critical to control the network load in order to maintain a high energy efficiency. Moreover, a two-tier small cell network with wireless backhaul can achieve a significant energy efficiency gain over a one-tier deployment, as long as the bandwidth division between radio access links and wireless backhaul is optimally designed.

Chapter 4 summarizes the main result, and provides several directions for future extension based on the frameworks we developed in this book.

References

1. J. Hoydis, K. Hosseini, S. ten Brink, and M. Debbah, "Making smart use of excess antennas: Massive MIMO, small cells, and TDD," *Bell Labs Tech. J.*, vol. 18, no. 2, pp. 5–21, Sep. 2013.
2. F. Rusek, D. Persson, B. K. Lau, E. G. Larsson, T. L. Marzetta, O. Edfors, and F. Tufvesson, "Scaling up MIMO: Opportunities and challenges with very large arrays," *IEEE Signal Process. Mag.*, vol. 30, no. 1, pp. 40–60, Oct. 2013.

3. E. Larsson, O. Edfors, F. Tufvesson, and T. Marzetta, "Massive MIMO for next generation wireless systems," *IEEE Commun. Mag.*, vol. 52, no. 2, pp. 186–195, Feb. 2014.
4. T. L. Marzetta, "Massive MIMO: An introduction," *Bell Labs Tech. J.*, vol. 20, pp. 11–22, 2015.
5. J. G. Andrews, H. Claussen, M. Dohler, S. Rangan, and M. C. Reed, "Femtocells: Past, present, and future," *IEEE J. Sel. Areas Commun.*, vol. 30, no. 3, pp. 497–508, Apr. 2012.
6. T. Q. S. Quek, G. de la Roche, I. Güvenç, and M. Kountouris, *Small cell networks: Deployment, PHY techniques, and resource management.* Cambridge University Press, 2013.
7. J. Hoydis, "Massive mimo and hetnets: Benefits and challenges," *Newcom# Summer School on Interference Management for Tomorrows Wireless Networks*, 2013.
8. V. Jungnickel, K. Manolakis, W. Zirwas, B. Panzner, V. Braun, M. Lossow, M. Sternad, R. Apel-fröjd, and T. Svensson, "The role of small cells, coordinated multipoint, and massive MIMO in 5G," *IEEE Commun. Mag.*, vol. 52, no. 5, pp. 44–51, May 2014.
9. E. Björnson, E. G. Larsson, and T. L. Marzetta, "Massive MIMO: Ten myths and one critical question," *IEEE Commun. Mag.*, vol. 54, no. 2, pp. 114–123, Feb. 2016.
10. X. Gao, O. Edfors, F. Rusek, and F. Tufvesson, "Massive MIMO in real propagation environments: Do all antennas contribute equally?" *IEEE Trans. Wireless Commun.*, vol. 63, no. 11, pp. 3917–3928, Nov. 2015.
11. T. L. Marzetta, "Noncooperative cellular wireless with unlimited numbers of base station antennas," *IEEE Trans. Wireless Commun.*, vol. 9, no. 11, pp. 3590–3600, Nov. 2010.
12. D. Gesbert, M. Kountouris, R. W. Heath Jr., C.-B. Chae, and T. Salzer, "Shifting the MIMO paradigm," *IEEE Signal Process. Mag.*, vol. 24, no. 5, pp. 36–46, 2007.
13. H. Q. Ngo, E. G. Larsson, and T. L. Marzetta, "Energy and spectral efficiency of very large multiuser mimo systems," *IEEE Trans. Commun.*, vol. 61, no. 4, pp. 1436–1449, Apr. 2013.
14. E. Björnson, M. Kountouris, and M. Debbah, "Massive MIMO and small cells: Improving energy efficiency by optimal soft-cell coordination," in *Proc. IEEE Int. Conf. on Telecommun. (ICT)*, Casablanca, Morocco, May 2013, pp. 1–5.
15. J. Jose, A. Ashikhmin, T. L. Marzetta, and S. Vishwanath, "Pilot contamination and precoding in multi-cell tdd systems," *IEEE Trans. Wireless Commun.*, vol. 10, no. 8, pp. 2640–2651, Aug. 2011.
16. X. Zhu, Z. Wang, C. Qian, L. Dai, J. Chen, S. Chen, and L. Hanzo, "Soft pilot reuse and multi-cell block diagonalization precoding for massive MIMO systems," *IEEE Trans. Veh. Technol.*, vol. PP, no. 99, pp. 1–1, 2015.
17. D. Gesbert, S. Hanly, H. Huang, S. S. Shitz, O. Simeone, and W. Yu, "Multi-cell MIMO cooperative networks: A new look at interference," *IEEE J. Sel. Areas Commun.*, vol. 28, no. 9, pp. 1380–1408, Dec. 2010.
18. E. Björnson, E. G. Larsson, and M. Debbah, "Massive MIMO for maximal spectral efficiency: How many users and pilots should be allocated?" *IEEE Trans. Wireless Commun.*, vol. 15, no. 2, pp. 1293–1308, Feb. 2016.
19. H. S. Dhillon, M. Kountouris, and J. G. Andrews, "Downlink MIMO hetnets: Modeling, ordering results and performance analysis," *IEEE Trans. Wireless Commun.*, vol. 12, no. 10, pp. 5208–5222, Oct. 2013.
20. Q. Ye, B. Rong, Y. Chen, M. Al-Shalash, C. Caramanis, and J. G. Andrews, "User association for load balancing in heterogeneous cellular networks," *IEEE Trans. Wireless Commun.*, vol. 12, no. 6, pp. 2706–2716, Jun. 2013.
21. A. J. Fehske, F. Richter, and G. P. Fettweis, "Energy efficiency improvements through micro sites in cellular mobile radio networks," in *Proc. IEEE Global Telecomm. Conf. Workshops*, Honolulu, HI, Dec. 2009, pp. 1–5.
22. Y. S. Soh, T. Q. S. Quek, M. Kountouris, and H. Shin, "Energy efficient heterogeneous cellular networks," *IEEE J. Sel. Areas Commun.*, vol. 31, no. 5, pp. 840–850, Apr. 2013.
23. M. Wildemeersch, T. Q. S. Quek, C. H. Slump, and A. Rabbachin, "Cognitive small cell networks: Energy efficiency and trade-offs," *IEEE Trans. Commun.*, vol. 61, no. 9, pp. 4016–4029, Sep. 2013.
24. H.-S. Jo, Y. J. Sang, P. Xia, and J. G. Andrews, "Heterogeneous cellular networks with flexible cell association: A comprehensive downlink SINR analysis," *IEEE Trans. Wireless Commun.*, vol. 11, no. 10, pp. 3484–3495, Oct. 2012.

25. S. Singh, H. S. Dhillon, and J. G. Andrews, "Offloading in heterogeneous networks: Modeling, analysis, and design insights," *IEEE Trans. Wireless Commun.*, vol. 12, no. 5, pp. 2484–2497, May 2013.
26. V. Jungnickel, K. Manolakis, S. Jaeckel, M. Lossow, P. Farkas, M. Schlosser, and V. Braun, "Backhaul requirements for inter-site cooperation in heterogeneous LTE-Advanced networks," in *Proc. IEEE Int. Conf. Commun.* Budapest, Hungary: IEEE, Jun. 2013, pp. 905–910.

Chapter 2
Massive MIMO for Interference Suppression: Cell-Edge Aware Zero Forcing

Abstract Ubiquitous high-speed coverage and seamless user experience are among the main targets of next generation wireless systems, and large antenna arrays have been identified as a technology candidate to achieve them. By exploiting the excess spatial degree of freedom from the large number of base station (BS) antennas, we propose a new scheme termed *cell-edge-aware* (CEA) zero forcing (ZF) precoder for coordinated beamforming in massive MIMO cellular network, which suppresses inter-cell interference at the most vulnerable user equipments (UEs). In this work, we combine the tools from random matrix theory and stochastic geometry to develop a framework that enables us to quantify the performance of CEA-ZF and compare that with a conventional *cell-edge-unaware* (CEU) ZF precoder in a network of random topology. Our analysis and simulations show that the proposed CEA-ZF precoder outperforms CEU-ZF precoding in terms of (i) increased aggregate per-cell data rate, (ii) higher coverage probability, and (iii) significantly larger 95 %-likely rate, the latter being the worst data rate that a UE can reasonably expect to receive when in range of the network. Results from our framework also reveal the importance of scheduling the optimal number of UEs per BS, and confirm the necessity to control the amount of pilot contamination received during the channel estimation phase.

2.1 Introduction

Supporting the ever increasing wireless throughput demand is the primary factor driving the industry and academia alike toward the fifth generation (5G) wireless systems. To attain a satisfied quality of service (QoS), 5G networks need not only to provide a large aggregate capacity, but also guarantee high worst-case rates for all UEs, including those located at the cell edge, i.e., close to interfering BSs [1–3]. New technologies are being introduced to improve the performance of cell-edge UEs from current levels. Equipping BSs with a large number of antennas, widely known as massive multiple-input multiple-output (MIMO), has emerged as one of the most promising solutions [4–6]. Besides the large diversity gain it brings along, spatial

© The Author(s) 2017   9
H.H. Yang and T.Q.S. Quek, *Massive MIMO Meets Small Cell*, SpringerBriefs
in Electrical and Computer Engineering, DOI 10.1007/978-3-319-43715-6_2

dimensions available at massive MIMO BSs can also be used to suppress interference at cell-edge UEs, thus shed light to new design aspects of inter-cell interference mitigation. To this end, we design and analyze a linear transmission scheme, termed *cell-edge-aware* (CEA) zero forcing (ZF) precoder, that significantly improves the data rate of UEs at the cell edge, as well as the overall network throughput.

2.1.1 Motivation and Related Work

A considerable amount of research has investigated the use of multi-cell joint signal processing for cell-edge performance improvement [7–9]. The common idea behind joint processing techniques is to organize BSs in clusters, where BSs lying in the same cluster share information on the data to be transmitted to all UEs in the cluster. Although this information allows BSs to coordinate their transmission and jointly serve all UEs with an improved system throughput, it comes at the cost of heavy signaling overhead and backhaul latency, which defy the purpose of its implementation [10].

As the benefits of joint processing are often outweighed by the increased latency and overhead, a more practical alternative to increase the cell-edge throughput can be found in coordinated beamforming, or *precoding*, schemes [11–13]. Under coordinated precoding, each BS acquires additional channel state information (CSI) of UEs in neighboring cells, but no data information is shared between the various BSs. The additional CSI can then be exploited to control the crosstalk generated at UEs in other cells, e.g., by using multiple BS antennas to steer the crosstalk toward the nullspace of the neighboring UEs. This approach is especially attractive for massive MIMO BSs, due to the abundance of spatial dimensions provided by the large antenna arrays [14].

Remarkable attempts to design and analyze a coordinated precoder for massive MIMO cellular networks are made in [15, 16]. The current paper differs from and generalizes the latter works in two key aspects

1. *Design*: Unlike [15, 16], where each BS suppresses the interference at all edge UEs in all neighboring cells, we specifically target those neighboring UEs close to the BS coverage area. Therefore, our precoder employs fewer spatial dimensions to mitigate inter-cell interference, leaving more degrees of freedom to each BS to better multiplex its own associated UEs [17].
2. *Analysis*: While [15, 16] assume a symmetric hexagonal cellular network, we consider a generalized model with random topology. Hexagonal models can lead to substantial performance overestimation, as demonstrated in [18, 19], whereas our analysis accounts for the randomness of practical cellular deployments.

2.1.2 Approach and Summary of Results

In this work, we propose a CEA-ZF precoder for massive MIMO cellular networks. Exploiting the excess spatial degrees of freedom made available by each massive MIMO BS, our CEA-ZF precoder suppress inter-cell interference at the cell-edge UEs in a distributed manner, thus boosting the received signal-to-interference ratio (SIR) of these most vulnerable of these most vulnerable UEs. In order to evaluate the performance of our scheme, we combine random matrix theory with stochastic geometry and analyze the coverage and rate performance of the proposed CEA-ZF precoder, as well as those of a conventional *cell-edge-unaware* (CEU) ZF precoder, in a general setting that accounts for random BS deployment and interference affecting both the channel estimation and data transmission phases. Our key results and contributions can be summarized as follows:

- We propose a new linear precoder for the downlink of massive MIMO cellular networks, which we denote as the CEA-ZF precoder, where some spatial dimensions are used to suppress inter-cell interference at the cell-edge neighboring UEs, and the remaining degrees of freedom are used to multiplex UEs within the cell. Our precoder works in a distributed manner, and it boosts network coverage and rate performance compared to CEU-ZF precoding.
- We develop a general framework to analyze the SIR distribution and coverage of massive MIMO cellular networks for both the proposed CEA-ZF and the CEU-ZF precoder. Our analysis is tractable and captures the effects of multi-antenna transmission, spatial multiplexing, path loss and small-scale fading, network load and BS deployment density, imperfect channel estimation, and random network topology.
- Through our analysis, which is validated via simulation results, we demonstrate that the proposed CEA-ZF precoder outperforms the CEU-ZF precoder in terms of aggregate per-cell data rate and coverage probability. Moreover, the CEA-ZF precoding guarantees a significantly larger 95 %-likely rate, i.e., it improves the worst data rate that any UE can expect to achieve.
- We quantify the effect of imperfect CSI, and reveal the importance of controlling the amount of pilot contamination received during the channel estimation phase, e.g., through smart pilot allocation schemes. We also study the system performance as a function of the network load, showing that it is beneficial to schedule the optimal number of UEs per BS.

The remainder of this chapter is organized as follows. We introduce the system model in Sect. 2.2. In Sect. 2.3, we analyze the SIR and network coverage under CEA-ZF and CEU-ZF precoding, also studying a special case that provides intuitions behind the gain attained by using different precoding schemes. We show the simulations that confirm the accuracy of our analysis as well as the numerical results to quantify the benefits of CEA-ZF precoding and obtain design insights in Sect. 2.4. This chapter is concluded in Sect. 2.5 (Main notations are summarized in Table 2.1).

Table 2.1 Notation summary

Notation	Definition
Φ_b; λ	PPP modeling the location of BSs; BS deployment density
N; K	Number of transmit antennas per BSs; number of scheduled UEs per BS
P_t; α	BS transmit power; path loss exponent
C_i; C_i^N	First-order Voronoi cell for BS i; cell neighborhood for BS i
$C_i^E = C_i \cup C_i^N$	Extended cell for BS i
r_{iik}; $r_{\bar{i}ik}$	Distance between UE k in cell i and its serving BS i and second closest BS $\bar{i}$, respectively
$\mathbf{x}_{ijk} \sim CN(0, \mathbf{I}_N)$	Small-scale fading between BS i and UE k in cell j
$\mathbf{w}_{u,ik}$; $\mathbf{w}_{a,ik}$	CEU-ZF and CEA-ZF precoding vector, respectively, at BS i for its UE k
M; F	Available number of pilots; pilot reuse factor
I_p	Interference during training phase
I_u; I_a	Interference during data transmission phase for CEU-ZF and CEA-ZF, respectively
τ_{ijk}	Standard deviation of CSI error between BS i and UE k in cell j
θ	SIR decoding threshold
$P_c(\theta)$; ρ_{95}	Coverage probability; 95 %-likely rate

2.2 System Model

We consider the downlink of a cellular network that consists of randomly deployed BSs, whose location follows a homogeneous Poisson point process (PPP) Φ_b of spatial density λ in the Euclidean plane. In this network, each BS transmits with power P_t and is equipped with a large number of antennas, N. Single-antenna UEs are distributed as a homogeneous PPP with sufficient high density on the plane, such that each BS has at least $K \leq N$ candidate UEs in its cell to serve. In light of its higher spectral efficiency, we consider spatial multiplexing at the BSs, where in each time–frequency resource block (RB) each BS simultaneously serves the K UEs in its cell [4].

We assume that UEs associate to the BS that provides the largest average received power. Due to the homogeneous nature of the network, this results in a distance-based association rule.[1] In this sense, the set of UE locations that are associated to BS i located at $z_i \in \mathbb{R}^2$ are defined by a classical Voronoi tessellation on the plane, denoted by V_i^1 and given by [22, 23]

$$V_i^1 = \left\{ \mathbf{z} \in \mathbb{R}^2 | \|\mathbf{z} - \mathbf{z}_i\| \leq \|\mathbf{z} - \mathbf{z}_k\|, \ \forall \mathbf{z}_k \in \Phi_b \setminus \mathbf{z}_i \right\}. \tag{2.1}$$

---

[1]Different association rules apply when transmit power or large-scale fading vary among BSs, resulting in a weighted Voronoi diagram [20, 21].

We note that the set V_i^1 contains all locations for which BS i is the closest. Such definition is identical to that of a traditional cell, thus we equivalently denote V_i^1 as C_i.

In order to identify the neighboring UEs for each BS, we find it useful to generalize the above definition of Voronoi cell to the second order. More precisely, the second-order Voronoi tessellation $V_{i,j}^2$ denotes the set of UE locations for which the BSs in $\mathbf{z}_i$ and $\mathbf{z}_j$ are the two closest, and it is given by [22, 23]

$$V_{i,j}^2 = \{\mathbf{z} \in \mathbb{R}^2 | \cap_{l \in \{i,j\}} \{\|\mathbf{z} - \mathbf{z}_l\| \leq \|\mathbf{z} - \mathbf{z}_k\|\}, \quad \forall \mathbf{z}_k \in \Phi_b \setminus \{\mathbf{z}_i, \mathbf{z}_j\}\}. \quad (2.2)$$

Using the second-order Voronoi tessellation, we can now define the notion of extended cell C_i^E for BS i, given by (i) all UEs for which BS i is the closest, and (ii) all UEs for which BS i is the second closest. The extended cell C_i^E is given by

$$C_i^E = \cup_j V_{i,j}^2, \quad \forall \mathbf{z}_j \in \Phi_b \setminus \mathbf{z}_i. \quad (2.3)$$

According to the above definition, each UE that lies in C_i^E sees BS i as either its closest or second closest BS. The extended cell C_i^E includes the UEs located in C_i that are served by BS i, as well as the neighboring UEs which are most vulnerable to interference generated by BS i. These UEs constitute the cell neighborhood for BS i, which we denote by C_i^N and define as follows:

$$C_i^N = C_i^E \setminus C_i. \quad (2.4)$$

Figure 2.1 illustrates the concepts of first-order and second-order Voronoi tessellation, cell neighborhood, and extended cell. Figure 2.1a shows a realization of first-order Voronoi tessellation, where each BS i covers a cell C_i. Figure 2.1b depicts the corresponding second-order Voronoi tessellation, where each pair of BSs (i, j)

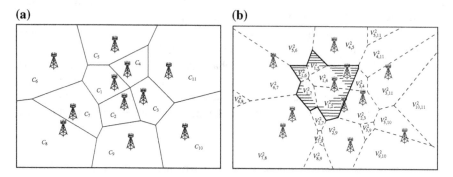

Fig. 2.1 Examples of (a) first-order and (b) second-order Voronoi tessellation. In (a), *solid lines* delimit first-order Voronoi cells C_i. In (b), *dashed lines* delimit second-order Voronoi cells $V_{i,j}^2$, *solid lines* delimit the extended cell C_1^E, and a *shadowed* region indicates the cell neighborhood C_1^N

identifies a region $V_{i,j}^2$ (delimited by dashed lines), such that UEs located in $V_{i,j}^2$ have BS i and BS j as their closest and second closest, or vice versa. Figure 2.1b also shows the extended cell C_1^E for BS 1 (delimited by solid lines), which is composed by the first-order cell C_1 and by the neighborhood C_1^N (shadowed region).

In the following, by using the concepts of extended cell and cell neighborhood, we propose a CEA-ZF precoding scheme where each BS not only spatially multiplexes the associated UEs in C_i, but also suppresses the interference caused at the most vulnerable neighboring UEs in C_i^N. We note that each BS i can easily obtain a list of UEs in C_i^N by means of reference signal received power (RSRP) estimation. In fact, downlink RSRP measurements for a list of neighboring BSs are periodically sent by each UE for handover purposes [24].

2.2.1 Channel Model and Estimation

In this network, we model the channels between any pair of antennas as independent and identically distributed (i.i.d.) and quasi-static, i.e., the channel is constant during a sufficiently long coherence block, and varies independently from block to block.[2] Moreover, we assume that each channel is narrowband and affected by two attenuation components, namely, small-scale Rayleigh fading, and large-scale path loss.[3] As such, the channel matrix from BS i to its K associated UEs can be written as

$$\mathbf{H}_i = \mathbf{R}_i^{\frac{1}{2}} \mathbf{X}_i, \tag{2.5}$$

where $\mathbf{R}_i = \mathrm{diag}\{r_{ii1}^{-\alpha}, \ldots, r_{iiK}^{-\alpha}\}$ is the path loss matrix, with r_{ijk} denoting the distance from the BS i to UE k in cell j, i.e., associated with BS j. The constant α represents the path loss exponent, whereas $\mathbf{X}_i = [\mathbf{x}_{ii1}, \ldots, \mathbf{x}_{iiK}]^H$ is the $K \times N$ fading matrix, where $\mathbf{x}_{ijk} \sim CN(0, \mathbf{I}_N)$ is the channel fading vector between BS i and UE k in cell j. Due to the interference-limited nature of massive MIMO cellular networks, we neglect the effect of thermal noise [4].

In order to simultaneously amplify the desired signal at the intended UEs and suppress interference at other UEs, each BS requires CSI from all the UEs it serves. This CSI is obtained during the training phase, where some RBs are used for the transmission of pilot signals. Since the number of pilots, i.e., the number of RBs allocated to the training phase, is limited, these pilots must be reused across cells. Pilot reuse implies that the estimate for the channel between a BS and one of its UEs

---

[2]Note that the results obtained through the machinery of random matrix theory can be modified to model transmit antenna correlation [25].

[3]For the sake of tractability, the analysis presented here does not consider shadowing. Note that the results involving large-system approximations can be adjusted to account for the presence of shadowing as in [20]. Moreover, a generalized gamma approximation can still be used under shadowing, since the channel attenuation at a given UE follows a Rayleigh distribution [21].

is contaminated by the channels between the BS and UEs in other cells which share the same pilot [4–6, 26].

Pilot contamination can be a limiting factor for the performance of massive MIMO. In order to mitigate such phenomenon, nonuniversal pilot reuse has been proposed, where neighboring cells use different sets of mutually orthogonal pilots [15, 19]. Under nonuniversal pilot reuse, the total set of available pilot sequences is divided into subgroups, and different subgroups are assigned to adjacent cells. For a pilot reuse factor F, the same subgroup of orthogonal pilot sequences is reused in every F cells.

We denote by $M = \kappa L$ the number of available orthogonal pilots, with L being the number of symbols that can be transmitted within a time–frequency coherence block, and κ being the fraction of symbols that are allocated for channel estimation. For a time-division duplexing (TDD) system with $L = 2 \times 10^4$ and $\kappa = 5\%$, there would be $M = 1000$ orthogonal pilots, and therefore a pilot reuse factor $F = 7$ would allow to estimate the channels of 142 UEs per cell [10].[4] As a general rule, a pilot reuse factor $F > 3$ is recommended in order to mitigate pilot contamination [28]. In this regard, we assume that there are sufficient pilot sequences to support a large enough pilot reuse factor, such that each BS can estimate the CSI of UEs in its own cell and in adjacent cells.

By using the MMSE criterion for pilot-based channel estimation, we can express the estimated small-scale fading $\hat{\mathbf{x}}_{ijk}$ between BS i and UE k in cell j as [17]

$$\mathbf{x}_{ijk} = \sqrt{1 - \tau_{ijk}^2}\hat{\mathbf{x}}_{ijk} + \tau_{ijk}\mathbf{q}_{ijk}, \tag{2.6}$$

where $\mathbf{q}_{ijk} \sim CN(\mathbf{0}, \mathbf{I}_N)$ is the normalized estimation error and τ_{ijk}^2 is the error variance, given by [9]

$$\tau_{ijk}^2 = \frac{1}{1 + \mathbb{E}[\gamma_{ijk}^{\mathrm{CSI}}] \cdot \frac{M}{FK}}. \tag{2.7}$$

In (2.7), $\gamma_{ijk}^{\mathrm{CSI}}$ is the SIR of the received pilot signal at the BS, given by

$$\gamma_{ijk}^{\mathrm{CSI}} = \frac{r_{ijk}^{-\alpha} \|\mathbf{x}_{ijk}\|^2}{\sum_{l \in \Phi_{\mathrm{P}}} r_{ilk}^{-\alpha} \|\mathbf{x}_{ilk}\|^2} \tag{2.8}$$

where Φ_{P} indicates the set of BSs that have their UEs reusing the same pilot as UE k in cell j and are thus generating pilot contamination. The estimated channel matrix at BS i can therefore be written as

---

[4]In a TDD system, downlink channels can be estimated through uplink pilots thanks to channel reciprocity. This makes the training time proportional to the number of UEs. A frequency-division duplexing (FDD) system requires a considerably longer training time, proportional to the number of BS antennas, and is therefore less suitable for massive MIMO [4, 27].

$$\hat{\mathbf{H}}_i = \mathbf{R}_i^{\frac{1}{2}} \hat{\mathbf{X}}_i. \tag{2.9}$$

2.2.2 Downlink Transmission

For the downlink transmission, we introduce two precoding schemes: (i) the conventional CEU-ZF precoder and (ii) the proposed CEA-ZF precoder.

2.2.2.1 Conventional CEU-ZF Precoding

With conventional zero forcing transmission, each BS i calculates the precoding vector to its UE k as [25]

$$\mathbf{w}_{\mathrm{u},ik} = \frac{1}{\sqrt{\zeta_{\mathrm{u},i}}} \left(\hat{\mathbf{H}}_i^{\mathrm{H}} \hat{\mathbf{H}}_i \right)^{-1} \hat{\mathbf{h}}_{iik}, \tag{2.10}$$

where $\hat{\mathbf{H}}_i = [\hat{\mathbf{h}}_{ii1}, \dots, \hat{\mathbf{h}}_{iiK}]$ is the estimated channel matrix, and $\zeta_{\mathrm{u},i} = \mathrm{tr}[(\hat{\mathbf{H}}_i^{\mathrm{H}} \hat{\mathbf{H}}_i)]$ is a power normalization factor.

Note that ZF precoding aims at mitigating intra-cell crosstalk caused by spatial multiplexing to attain better SIR, where each BS projects signal for each intended UE onto the null space of other UEs receiving service in the same cell. In a multiuser MIMO system, it has been shown that ZF outperform maximum ratio transmission (MRT) in terms of per-cell sum-rate [15]. When the system dimensions make the ZF matrix inversion in (2.10) computationally expensive, a simpler truncated polynomial expansion can be employed with similar performance [29].

2.2.2.2 Proposed CEA-ZF Precoding

Unlike CEU-ZF precoding, where all spatial dimensions available at each BS i are used to multiplex UEs within cell C_i, the proposed CEA-ZF precoder exploits some spatial dimensions to suppress interference at the most vulnerable UEs, i.e., those lying in the BS's cell neighborhood C_i^{N}. In this fashion, BS i sacrifices some of its intra-cell power gain while trying to be a good neighbor who reduces the inter-cell interference to each of its adjacent cells. If such interference suppression is performed by all BSs in a distributed manner, the cell-edge performance as well as the overall data rate of the network can be improved. An illustration of the basic features of CEA-ZF is given in Fig. 2.2, where a multi-antenna BS spatially multiplexes its in-cell UEs while simultaneously suppressing interference at its neighboring UEs.

For BS i, we denote by K' the number of UEs lying in the cell neighborhood, where we omit the subscript i for notational convenience. We note that K' indicates the number of UEs for which BS i is the second closest, i.e., the number of neighboring

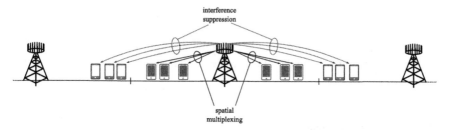

Fig. 2.2 Illustration of CEA-ZF, where a BS performs spatial multiplexing for in-cell UEs (*blue*) and interference suppression for neighboring UEs (*red*)

UEs for BS i. The proposed CEA-ZF precoder between BS i and UE k in cell i is then given by

$$\mathbf{w}_{\mathrm{a},ik} = \frac{1}{\sqrt{\zeta_{\mathrm{a},i}}} \left(\sum_{l=1}^{K} \hat{\mathbf{h}}_{iil} \hat{\mathbf{h}}_{iil}^{\mathrm{H}} + \sum_{l=1}^{K'} \hat{\mathbf{h}}_{i\bar{i}l} \hat{\mathbf{h}}_{i\bar{i}l}^{\mathrm{H}} \right)^{-1} \hat{\mathbf{h}}_{iik}, \qquad (2.11)$$

with $\hat{\mathbf{h}}_{i\bar{i}l}$ denoting the estimated channel between BS i and the l-th neighboring UE, where the notation $\bar{i}$ indicates that BS i is the second closest BS for that particular UE. The constant $\zeta_{\mathrm{a},i}$ is chosen as an average power normalization factor, given by $\zeta_{\mathrm{a},i} = \sum_{k=1}^{K} \|\mathbf{w}_{\mathrm{a},ik}\|^2$.

It should be noted that the CEA-ZF precoder in (2.11) can be seen as a generalization of the two-cell precoder proposed in [11] to a nonsymmetric and non-pairwise scenario.

2.3 Coverage Analysis

In this section, we analyze the downlink SIR coverage of a massive MIMO cellular network with conventional CEU-ZF precoding and the proposed CEA-ZF precoding, and we provide a special case that helps us to grasp intuitions about the performance gain attained from CEA-ZF.

2.3.1 Preliminaries

2.3.1.1 Coverage Probability

In our analysis, the performance metric of interest is the coverage probability, defined as the probability that the received SIR γ at a generic UE is above a given threshold θ, i.e.,

$$P_c(\theta) = \mathbb{P}(\gamma \geq \theta), \quad \theta > 0. \tag{2.12}$$

We note that the coverage probability $P_c(\theta)$ provides information on the SIR distribution across the network, and it allows to evaluate the data rate performance at the cell edge.

2.3.1.2 SIR at a Typical UE

By applying Slivnyak's theorem to the stationary PPP of BSs, it is sufficient to evaluate the SIR of a typical UE at the origin [30]. In the following, we denote as *typical* the UE k associated with BS i, with a received signal given by

$$y_{ik} = P_t\mathbf{h}_{iik}\mathbf{w}_{ik}s_{ik} + \sum_{l=1,l\neq k}^{K} P_t\mathbf{h}_{iik}\mathbf{w}_{il}s_{il} + \sum_{j\neq i}^{\infty}\sum_{l=1}^{K} P_t\mathbf{h}_{jik}\mathbf{w}_{jl}s_{jl}, \tag{2.13}$$

where $\mathbf{w}_{ik}$ is the normalized precoding vector from the serving BS i to the typical UE, and s_{ik} is the corresponding unit-power signal, i.e., $\mathbb{E}\left[|s_{ik}|^2\right] = 1$. The vector $\mathbf{w}_{ik}$ can take different forms, depending on the precoding scheme employed. The SIR at the typical UE can be written as

$$\gamma_{ik} = \frac{|\mathbf{h}_{iik}\mathbf{w}_{ik}|^2}{\sum_{l=1,l\neq k}^{K}|\mathbf{h}_{iik}\mathbf{w}_{il}|^2 + I}, \tag{2.14}$$

where the first summation in the denominator represents the intra-cell interference, while I denotes the aggregate out-of-cell interference. The latter is given by

$$I = \sum_{j\neq i} \frac{g_{jik}}{r_{jik}^{\alpha}}, \tag{2.15}$$

where $g_{jik} = \sum_{l=1}^{K}|\mathbf{x}_{jik}\mathbf{w}_{jl}|^2$ is the effective small-scale fading from interfering BS j to UE k in cell i. From results in [31], we note that if the precoding vectors $\{\mathbf{w}_{jl}\}_{l=1}^{K}$ at BS j are mutually independent and satisfy $\sum_{l=1}^{K}\|\mathbf{w}_{jl}\|^2 = 1$, the effective channel fading is distributed as $g_{jik} \sim \Gamma(K, 1/K)$.

2.3.1.3 CSI Error

Under sufficient nonuniversal pilot reuse, a generic BS i can estimate the channels of all in-cell UEs as well as the channels of neighboring UEs. From (2.7) and (2.8), the CSI error variance for an in-cell UE and for a neighboring UE can be, respectively, written as

$$\tau^2 = \frac{1}{1 + \frac{t^{-\alpha}}{I_p} \cdot \frac{M}{FK}},\qquad(2.16)$$

$$\bar{\tau}^2 = \frac{1}{1 + \frac{s^{-\alpha}}{I_p} \cdot \frac{M}{FK}},\qquad(2.17)$$

where t and s denote the distance between a typical UE and its closest and second closest BS, respectively, and I_p is the pilot interference received during the training phase.

Under reuse factor F, clusters of F adjacent cells choose different subgroups of pilot sequences and do not cause interference, i.e., pilot contamination to each other. Therefore, each BS receives pilot contamination only from UEs lying outside the cluster of F cells, whose mean area can be calculated as F/λ [30]. This area can be approximated with a circle $B(0, R_e)$ of radius $R_e = \sqrt{F/(\lambda\pi)}$ [32], yielding the following mean interference

$$\mathbb{E}\left[I_p\right] = \mathbb{E}\left[\sum_{x \in \Phi_p \cap B^c(0, R_e)} h_{x,o}\|x\|^{-\alpha}\right]$$
$$= \frac{2\left(\lambda\pi/F\right)^{\frac{\alpha}{2}}}{\alpha - 2},\qquad(2.18)$$

where $B^c(0, R_e)$ denotes the complement set of $B(0, R_e)$. By approximating the interference I_p with its mean [9] and by substituting (2.18) into (2.16) and (2.17), the CSI error can be expressed as a function of t and s as follows:

$$\tau^2 \approx \frac{1}{1 + \frac{M(\alpha-2)F^{\frac{\alpha}{2}-1}}{2K(\lambda\pi)^{\frac{\alpha}{2}}t^{\alpha}}},\qquad(2.19)$$

$$\bar{\tau}^2 \approx \frac{1}{1 + \frac{M(\alpha-2)F^{\frac{\alpha}{2}-1}}{2K(\lambda\pi)^{\frac{\alpha}{2}}s^{\alpha}}}.\qquad(2.20)$$

2.3.2 Coverage Probability Under CEU-ZF

We now derive the coverage probability under CEU-ZF precoding. An approximation of the conditional SIR under CEU-ZF can be obtained in the large-system regime as follows.

Lemma 2.1 *Conditioned on the out-of-cell interference I_u and the intra-cell distance r_{iil}, $l \in \{1, \ldots, K\}$, when $K, N \to \infty$ with $\beta = K/N < 1$, the SIR achieved by CEU-ZF precoding converges almost surely to the following quantity*

$$\gamma_{u,ik} \rightarrow \frac{\left(1 - \tau_{iik}^2\right)\left(1 - \beta\right)N}{\left(\tau_{iik}^2 r_{iik}^{-\alpha} + I_u\right)\left(r_{iik}^\alpha + R_k\right)},$$

(2.21)

where R_k is given by

$$R_k = \sum_{l=1, l \neq k}^{K} r_{iil}^\alpha.$$

(2.22)

Proof See Appendix section "Proof of Lemma 2.1".

Deriving the coverage probability requires knowledge of the distribution of R_k, which is the sum of $(K-1)$ i.i.d. random variables (r.v.s) r_{iii}^α. The distance r_{lii} is a r.v. that follows a Rayleigh distribution $f_c(r)$, given by [33]

$$f_c(r) = 2\pi\lambda r e^{-\lambda\pi r^2}.$$

(2.23)

It can then be shown that r_{lii}^α follows a Weibull distribution with shape and scale parameters $2/\alpha$ and $(\lambda\pi)^{-\frac{\alpha}{2}}$, respectively [34]. As such, the distribution of R_k can be approximated by a generalized Gamma distribution as follows [35].

Assumption 1 The probability density function (pdf) $f_{R_k}(r)$ and cumulative density function (CDF) $F_{R_k}(r)$ of the r.v. R_k can be approximated as follows:

$$f_{R_k}(r) \approx \frac{\eta\mu^\mu r^{\eta\mu-1}}{\Omega^\mu\Gamma(\mu)} \exp\left(-\frac{\mu r^\eta}{\Omega}\right),$$

(2.24)

$$F_{R_k}(r) \approx \frac{1}{\Gamma(\mu)}\Gamma\left(\mu, \frac{\mu r^\eta}{\Omega}\right),$$

(2.25)

where $\Gamma(s, x) = \int_0^x t^{s-1}e^{-t}dt$ is the lower incomplete gamma function, and $\Omega = \mathbb{E}[R_k^\eta]$ is a scale parameter, given by

$$\Omega = \left[\frac{\mu^{\frac{1}{\eta}}\Gamma(\mu)\,\mathbb{E}[R_k]}{\Gamma\left(\mu + \frac{1}{\eta}\right)}\right]^\eta$$

(2.26)

while μ and η are solutions of the following equations:

$$\frac{\Gamma^2\left(\mu + \frac{1}{\eta}\right)}{\Gamma(\mu)\,\Gamma\left(\mu + \frac{2}{\eta}\right) - \Gamma^2\left(\mu + \frac{1}{\eta}\right)} = \frac{\mathbb{E}^2[R_k]}{\mathbb{E}[R_k^2] - \mathbb{E}^2[R_k]},$$

(2.27)

$$\frac{\Gamma^2\left(\mu + \frac{2}{\eta}\right)}{\Gamma(\mu)\,\Gamma\left(\mu + \frac{4}{\eta}\right) - \Gamma^2\left(\mu + \frac{2}{\eta}\right)} = \frac{\mathbb{E}^2[R_k^2]}{\mathbb{E}[R_k^4] - \mathbb{E}^2[R_k^2]}.$$

(2.28)

The quantities $\mathbb{E}[R_k]$, $\mathbb{E}[R_k^2]$, and $\mathbb{E}[R_k^4]$ are the first, second, and fourth moment of the r.v. R_k, respectively, and can be calculated as

$$\mathbb{E}[R_k] = \frac{K-1}{(\lambda\pi)^{\frac{\alpha}{2}}} \Gamma\left(1+\frac{\alpha}{2}\right), \tag{2.29}$$

$$\mathbb{E}[R_k^2] = \frac{K-1}{(\lambda\pi)^{\alpha}} \left[\Gamma(1+\alpha) + (K-2)\,\Gamma^2\left(1+\frac{\alpha}{2}\right)\right], \tag{2.30}$$

$$\mathbb{E}[R_k^4] = \frac{K-1}{(\lambda\pi)^{2\alpha}} \left[(K-2)\,(K-3)\,(K-4)\,\Gamma^4\left(1+\frac{\alpha}{2}\right)\right.$$

$$+ 3(K-2)\,\Gamma^2(1+\alpha) + 4\,(K-2)\,\Gamma\left(1+\frac{\alpha}{2}\right)\Gamma\left(1+\frac{3\alpha}{2}\right)$$

$$\left.+ \Gamma(1+2\alpha) + 6\,(K-2)\,(K-3)\,\Gamma(1+\alpha)\,\Gamma^2\left(1+\frac{\alpha}{2}\right)\right]. \tag{2.31}$$

By using the approximated distribution of R_k, we can now obtain the coverage probability of a massive MIMO cellular network under CEU-ZF.

Theorem 2.1 *The coverage probability of a massive MIMO cellular network under CEU-ZF precoding can be approximated as*

$$\mathbb{P}\left(\gamma_{u,ik} \geq \theta\right) \approx \frac{1}{\Gamma(\mu)} \int_0^\infty \Gamma\left(\mu, \frac{\mu}{\Omega}\left[\frac{(1-\tau^2)(1-\beta)\,Nr^\alpha}{\theta\left(\tau^2+\frac{2\pi\lambda r^2}{\alpha-2}\right)} - r^\alpha\right]^\eta\right) f_c(r)dr, \tag{2.32}$$

where τ^2 is given in (2.19), and $f_c(r)$ is given by (2.23).

Proof See Appendix section "Proof of Theorem 2.1".

Here, it should be noted that although coverage probability of a multiuser MIMO cellular network with conventional CEU-ZF precoding has also been derived in [31, 36, 37], the result in (2.32) provides an approximation that involves only one integration and is therefore easier to be evaluated. The accuracy of this approximation will be verified in Fig. 2.4.

2.3.3 Coverage Probability Under CEA-ZF

We now derive the coverage probability under the proposed CEA-ZF precoder. Similar to the CEU-ZF, an approximation of the conditional SIR under CEA-ZF can be obtained in the large-system regime as follows.

Lemma 2.2 *Conditioned on the out-of-cell interference I_a, the intra-cell distance r_{iil} with $l \in \{1, \ldots, K\}$, the distance r_{iik} between the typical UE and its second* *closest BS, and the standard deviation $\tau_{\bar{i}ik}$ of the corresponding CSI error, when $K, N \to \infty$ with $\beta = K/N < 1$ and $\beta' = K'/N < 1$, the SIR of CEA-ZF converges almost surely to a quantity given by*

$$\gamma_{a,ki} \to \frac{\left(1 - \tau_{iik}^2\right)\left(1 - \beta - \beta'\right)N}{\left(\tau_{iik}^2 r_{iik}^{-\alpha} + \tau_{\bar{i}ik}^2 r_{\bar{i}ik}^{-\alpha} + I_a\right)\left(r_{iik}^\alpha + R_k\right)} \tag{2.33}$$

Proof See Appendix section "Proof of Lemma 2.2".

Using the above results, we are now able to derive the coverage probability under CEA-ZF precoding.

Theorem 2.2 *The coverage probability of a massive MIMO cellular network under CEA-ZF can be obtained as*

$$\mathbb{P}\left(\gamma_{a,ik} \geq \theta\right) \approx \int\limits_0^\infty \int\limits_t^\infty \Gamma\left(\mu, \frac{\mu}{\Omega}\left[\frac{\left(1 - \tau^2\right)\left(1 - 2\beta\right)N s^\alpha}{\theta\left(\bar{\tau}^2 + \tau^2 \frac{s^\alpha}{t^\alpha} + \frac{2\pi\lambda s^2}{\alpha - 2}\right)} - t^\alpha\right]^\eta\right) \frac{f_{s|c}(s, t) f_c(t)}{\Gamma(\mu)} \, ds \, dt, \tag{2.34}$$

where τ^2 and $\bar{\tau}^2$ are given in (2.19) and (2.20), respectively, and

$$f_{s|c}(s, t) = 2\pi\lambda s e^{-\lambda\pi\left(s^2 - t^2\right)}. \tag{2.35}$$

Proof See Appendix section "Proof of Theorem 2.2".

Equations (2.32) and (2.34) quantify how some of the key features of a cellular network, i.e., deployment strategy, interference, and impairments in the channel estimation phase, affect the coverage probability provided by massive MIMO BSs. These equations will be used in our numerical study in the following section, which investigates the impact of different system parameters on the network performance. Before going to that directly, we provide a special case that helps us grasp the first order understanding from network parameter impacts, stated in the following corollary.

Corollary 2.1 *In the case that $\tau^2 = \bar{\tau}^2 = 0$, when $\alpha = 4$ and $N \to \infty$, the coverage probability under CEU-ZF and CEA-ZF converges respectively as follows:*

$$\mathbb{P}\left(\gamma_{u,ik} \geq \theta\right) \to 1 - \frac{2(K - 1)\theta}{(1 - \beta)N} \approx 1 - \frac{2\beta\theta}{1 - \beta}, \tag{2.36}$$

$$\mathbb{P}\left(\gamma_{a,ik} \geq \theta\right) \to 1 - \frac{4(K - 1)(K + 4)\theta^2}{(1 - 2\beta)^2 N^2} \approx 1 - \left(\frac{2\beta\theta}{1 - 2\beta}\right)^2. \tag{2.37}$$

Proof See Appendix section "Proof of Corollary 2.1".

Several conclusions can be drawn from the above corollary, we state them as follows:

- CEA-ZF outperforms CEU-ZF in terms of coverage probability only when $\beta < \frac{1}{2} - \frac{\sqrt{\theta^2+2\theta}}{4+2\theta}$, or equivalently, $\frac{2(4+\theta)K}{2+\theta-\sqrt{\theta^2+2\theta}} < N$. This observation implies that CEA-ZF requires sufficient spatial dimensions to attain better performance.
- When BS antenna number N grows to be large, the coverage probability under CEU-ZF and CEA-ZF increases as $1/N$ and $1/N^2$, respectively. This observation demonstrates that CEA-ZF is more effective precoding scheme in massive MIMO cellular networks.

2.4 Numerical Results and Discussion

In this section, we first show simulation results that confirm the accuracy of our analytical framework. After that, we provide numerical results to show the performance gain attained by the proposed CEA-ZF precoder, and we discuss how data rate and coverage are affected by the number of scheduled UEs and the channel estimation accuracy.

2.4.1 Simulation Validations

Unless differently specified, we use the following parameters for path loss exponent, BS density, and number of UE, respectively: $\alpha = 3.8$, $\lambda = 10^{-6}$, and $K = 10$.

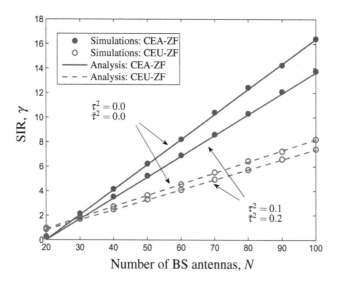

Fig. 2.3 Downlink SIR for CEU-ZF and CEA-ZF versus number of BS antennas, for scenarios with and without CSI error

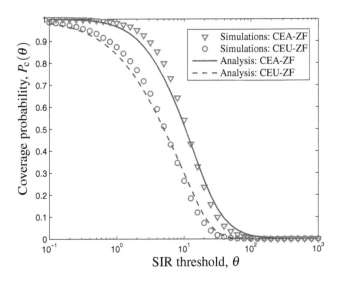

Fig. 2.4 Coverage probability under CEU-ZF and CEA-ZF precoding

In Fig. 2.3a, we depict the downlink SIR achieved by a typical UE of a massive MIMO cellular network as a function of the number of BS antennas N, under different precoding schemes and transmit CSI errors. The figure shows a negligible difference between simulations and analytical results, which confirms the accuracy of Lemmas 2.1 and 2.2. We also note that the SIR values obtained for ZF precoding are consistent with the ones obtained in [17].

Figure 2.4 compares the simulated coverage probability to the analytical results derived in Theorems 2.1 and 2.2. The coverage probability is plotted versus the SIR threshold at the typical UE. The figure shows that analytical results and simulations fairly well match and follow the same trend, thus confirming the accuracy of the theorems.

2.4.2 Numerical Results

Unless otherwise stated, the following system parameters will be used: BS deployment density $\lambda = 10^{-6}$, number of scheduled UEs per cell $K = 20$, path loss exponent $\alpha = 3.8$, and pilot reuse factor $F = 7$. We note that neglecting the training time does not affect the fairness of our performance comparison between CEU-ZF and CEA-ZF precoding.

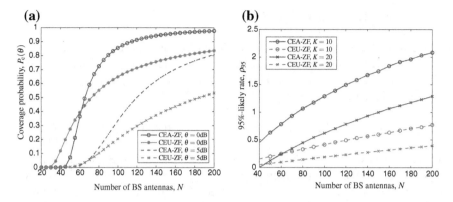

Fig. 2.5 Impact of antenna number on network performance: **a** coverage probability versus number of BS antennas for K = 20 scheduled UEs per cell, under CEU-ZF and CEA-ZF precoding. **b** 95 %-likely rate versus number of BS antennas, with CEU-ZF and CEA-ZF precoding

2.4.2.1 Impact of Large Antenna Array

In Fig. 2.5, we show the effect of BS antenna number N on the network performance. Figure 2.5a depicts the coverage probability under CEU-ZF and CEA-ZF precoding as a function of the number of BS antennas N, for two different SIR thresholds θ. The following can be observed: (i) with a sufficient number of BS antennas, i.e., $N > 3K$, CEA-ZF outperforms CEU-ZF precoding, and (ii) CEA-ZF requires significantly fewer antennas to achieve high coverage probabilities compared to CEU-ZF precoding.

In Fig. 2.5b, we compare the 95 %-likely rate under the two precoding schemes. The 95 %-likely rate (denoted by ρ_{95}) is defined as the rate achievable by at least 95 % of the UEs in the network, and it can be regarded as the worst rate any scheduled UE may expect to receive when located at the cell edge [3, 4]. While the 95 %-likely rates of both CEU-ZF and CEA-ZF precoding benefit from a larger number of BS antennas N, the proposed CEA-ZF precoder achieves a significantly larger 95 %-likely rate compared to conventional CEU-ZF, and the gain increases as N grows.

In summary, the proposed CEA-ZF precoder outperforms conventional CEU-ZF precoding from several perspectives. CEA-ZF provides better coverage than CEU-ZF, especially in the massive MIMO regime, i.e., when BSs are equipped with a large number of antennas, N. While CEA-ZF can attain high coverage probability with reasonable values of N, a significantly larger number of antennas is required to achieve the same coverage under CEU-ZF precoding. The proposed CEA-ZF precoder also achieves a larger sum-rate per cell, and a significantly larger 95 %-likely rate. The latter is especially important, being the worst data rate that any scheduled UE can expect to receive.

2.4.2.2   Impact of Imperfect CSI

We now study the impact of the channel estimation error on coverage and edge rates. To this end, we vary the CSI error variance τ^2 and $\bar{\tau}^2$ at in-cell UEs and neighboring UEs, respectively, while keeping their ratio constant as $\mathbb{E}[\bar{\tau}^2]/\mathbb{E}[\tau^2] = 1.8$. In the following, we set the SIR threshold as $\theta = 0\,\text{dB}$ and number of BS antennas as $N = 100$.

Figure 2.6a shows the coverage probability as a function of the CSI error for various values of scheduled UEs per cell. Although the presence of a CSI error degrades the coverage probability of both CEU-ZF and CEA-ZF precoding, it can be seen that CEA-ZF significantly outperforms conventional CEU-ZF for low-to-moderate values of the CSI error. Under a large CSI error, CEA-ZF still performs as well as CEU-ZF as long as the number of scheduled UEs per cell is controlled, e.g., $K = 10$ or $K = 20$ in the figure.

Figure 2.6b depicts the 95 %-likely rate as a function of the CSI error variance for CEU-ZF and CEA-ZF precoding. Once again, CEA-ZF significantly outperforms conventional CEU-ZF for low-to-moderate values of the CSI error, while the 95 %-likely rates of both precoders degrade and achieve similar values under very poor CSI quality, i.e., large values of τ^2 and $\bar{\tau}^2$.

As expected, pilot contamination can negatively affect the achievable data rates by degrading the quality of the CSI available at the BSs. In the presence of very large channel estimation errors, the performance of CEA-ZF precoding degrades and converges to the one of conventional CEU-ZF precoding. In fact, the cell-edge suppression mechanism employed by CEA-ZF relies on the accuracy of the measured channels, and the promised gains in terms of coverage and 95 %-likely rate cannot be achieved. It is therefore desirable to control the amount of pilot contamination received during the channel estimation phase, for example by designing appropriate pilot allocation schemes.

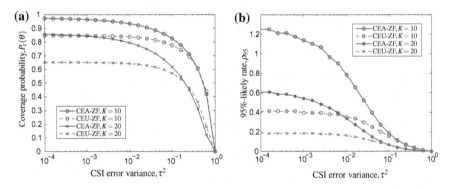

Fig. 2.6 Comparison of coverage probability and 95 %-likely rate with CSI error variance. In (**a**), Coverage probability plotted as a function of CSI error variance. In (**b**), 95 %-likely rate plotted as a function of CSI error variance

2.5 Conclusion

In this chapter, we proposed the CEA-ZF precoder that exploits the excess spatial degrees of freedom available at massive MIMO BSs to suppress inter-cell interference at the most vulnerable UEs in the network. The CEA-ZF specifically targets those neighboring UEs close to the BS coverage area, thus requiring fewer spatial dimensions to mitigate inter-cell interference, and leaving more dimensions for intra-cell spatial multiplexing. Moreover, it can be implemented in a distributed fashion.

In order to model practical deployments, we analyzed the performance of CEA-ZF and conventional CEU-ZF precoding in a random asymmetric cellular network. By using CEA-ZF precoder, we showed that a better network coverage is attainable. More importantly, the 95 %-likely rate, namely, the minimum data rate that any UE can expect to achieve, is significantly improved. The latter is of particular interest, given the ambitious edge rate requirements set for 5G, which aims at uninterrupted user experience. Our study also quantified the impact of imperfect CSI, confirming the importance of controlling the amount of pilot contamination during the channel estimation phase.

Appendix

Proof of Lemma 2.1

We treat the out-of-cell interference as noise, then as $K, N \to \infty$ with $\beta = K/N < 1$, the SIR converges to [25]

$$\gamma_{u,ik} \to \frac{1 - \tau_{iik}^2}{\Upsilon \cdot r_{iik}^{-\alpha} \tau_{iik}^2 + \Psi \cdot I_u}, \quad a.s. \tag{2.38}$$

where Υ and Ψ are given by

$$\Upsilon = \frac{1}{\phi} \frac{1}{N} \mathrm{tr} \mathbf{R}_i^{-1}, \tag{2.39}$$

$$\Psi = \frac{\psi}{\frac{N}{K}\phi^2 - \psi} \frac{1}{K} \frac{1}{N} \mathrm{tr} \mathbf{R}_i^{-1} \tag{2.40}$$

and ψ and ϕ are given, respectively, by

$$\psi = \frac{1}{N} \mathrm{tr} \left(\mathbf{I}_N + \frac{K}{N} \frac{1}{\phi} \mathbf{I}_N \right)^{-2} = \frac{1}{\left(1 + \frac{\beta}{\phi}\right)^2}, \tag{2.41}$$

$$\phi = \frac{1}{N} \mathrm{tr}\left[\left(1+\frac{\beta}{\phi}\right)\mathbf{I}_N\right]^{-1} = \frac{1}{1+\frac{\beta}{\phi}}. \tag{2.42}$$

By solving (2.41) and (2.42), we obtain $\phi = 1 - \beta$ and $\psi = \phi^2$. By substituting ψ and ϕ in (2.39) and (2.40), respectively, the following holds

$$\Upsilon = \Psi = \frac{1}{N}\frac{1}{1-\beta}\sum_{l=1}^{K} r_{iil}^{\alpha}. \tag{2.43}$$

Lemma 2.1 then follows by substituting (2.43) into (2.38).

Proof of Theorem 2.1

Consider a typical UE k in cell i and located at the origin. We denote the distance between the typical UE and its serving BS i as $r_{iik} = t$. As such, we can approximate the out-of-cell interference I_u by its mean, which can be computed as

$$\mathbb{E}\left[I_u\right] = \mathbb{E}\left[\sum_{x\in\Phi_b\backslash i}\frac{g_{xik}}{\|x\|^{\alpha}}\right]$$
$$\overset{(a)}{=}\int_t^{\infty} r^{-\alpha} r dr = \frac{2\pi\lambda t^{-(\alpha-2)}}{\alpha-2}, \tag{2.44}$$

where (a) is obtained by applying the Campbell's theorem [30]. By substituting (2.44) into (2.21), the conditional SIR received at the typical UE can be approximated as

$$\gamma_{u,ik} \approx \frac{\left(1-\tau_{iik}^2\right)(1-\beta)N}{\left(\tau_{iik}^2 t^{-\alpha} + \frac{2\pi\lambda t^{-(\alpha-2)}}{\alpha-2}\right)(t^{\alpha}+R_k)}. \tag{2.45}$$

The coverage probability can then be calculated as

$$\mathbb{P}\left(\gamma_{u,ik}\geq\theta\right) \approx \mathbb{E}\left[\mathbb{P}\left(R_k \leq \frac{\left(1-\tau_{iik}^2\right)(1-\beta)Nt^{\alpha}}{\left(\tau_{iik}^2+\frac{2\pi\lambda t^2}{\alpha-2}\right)\theta} - t^{\alpha}\,\middle|\,r_{iik}=t\right)\right]$$
$$= \frac{1}{\Gamma(\mu)}\mathbb{E}\left[\Gamma\left(\mu,\frac{\mu}{\Omega}\left[\frac{(1-\tau^2)(1-\beta)Nt^{\alpha}}{\theta\left(\tau^2+\frac{2\pi\lambda t^2}{\alpha-2}\right)}-t^{\alpha}\right]^{\eta}\right)\right]. \tag{2.46}$$

Since the pdf of t has been given in (2.23), Theorem 2.1 then follows from (2.46) by deconditioning t.

Proof of Lemma 2.2

We start with the minimum-mean-square-error (MMSE) version of the CEA-ZF precoder, which includes a regularization term $\rho \mathbf{I}_N$, given by

$$\tilde{\mathbf{w}}_{\mathrm{a},ik} = \frac{1}{\sqrt{\tilde{\zeta}_{\mathrm{a},i}}} \left(\rho N \mathbf{I}_N + \sum_{l=1}^{K} \hat{\mathbf{h}}_{iil} \hat{\mathbf{h}}_{iil}^{\mathrm{H}} + \sum_{l=1}^{K'} \hat{\mathbf{h}}_{i\bar{i}l} \hat{\mathbf{h}}_{i\bar{i}l}^{\mathrm{H}} \right)^{-1} \hat{\mathbf{h}}_{iik}, \qquad (2.47)$$

where $\tilde{\zeta}_{\mathrm{a},i} = \sum_{k=1}^{K} \left\| \tilde{\mathbf{w}}_{\mathrm{a},ik} \right\|^2$. The SIR at the typical UE can be written as

$$\gamma_{\mathrm{a},ik} = \frac{\left| \mathbf{h}_{iik}^{\mathrm{H}} \tilde{\mathbf{w}}_{\mathrm{a},ik} \right|^2}{\sum_{l \neq k} \left| \mathbf{h}_{iik}^{\mathrm{H}} \tilde{\mathbf{w}}_{\mathrm{a},il} \right|^2 + \sum_{l=1}^{K} \left| \mathbf{h}_{\bar{i}ik}^{\mathrm{H}} \tilde{\mathbf{w}}_{\mathrm{a},\bar{i}l} \right|^2 + I_{\mathrm{a}}}. \qquad (2.48)$$

Substituting (2.9) into the numerator of (2.48), and using the matrix inversion lemma and the rank-1 permutation lemma [25], the received signal power converges to the following limit in the large-system regime:

$$\left| \hat{\mathbf{h}}_{iik}^{\mathrm{H}} \tilde{\mathbf{w}}_{\mathrm{a},ik} \right|^2 \rightarrow \frac{1}{\zeta_{\mathrm{a},i}} \frac{\left(r_{iik}^{-\alpha} \Lambda_i \right)^2}{\left(1 + r_{iik}^{-\alpha} \Lambda_i \right)^2}, \qquad \text{as } N \rightarrow \infty \qquad (2.49)$$

where Λ_i is the solution of the following fix point equation [38]

$$\Lambda_i = \frac{1}{\rho + \frac{1}{N} \sum_{l=1}^{K} \frac{r_{iil}^{-\alpha}}{1 + \Lambda_i r_{iil}^{-\alpha}} + \frac{1}{N} \sum_{l=1}^{K'} \frac{r_{i\bar{i}l}^{-\alpha}}{1 + \Lambda_i r_{i\bar{i}l}^{-\alpha}}}. \qquad (2.50)$$

We next deal with the first two summations in the denominator of (2.48), which are the intra-cell interference from the serving BS and the inter-cell interference from the second closest interfering BS. Similarly, by using the rank-1 permutation again, the large-system limit for these interference read as

$$\sum_{l \neq k} \left| \mathbf{h}_{iik}^{\mathrm{H}} \tilde{\mathbf{w}}_{\mathrm{a},il} \right|^2 = \sum_{l} \frac{\frac{1}{\zeta_{\mathrm{a},i}} \frac{r_{iil}^{-\alpha}}{N} \left(-\frac{\partial \Lambda_i}{\partial \rho} \right)}{\left(1 + r_{iil}^{-\alpha} \Lambda_i \right)^2} \left[\frac{\left(1 - \tau_{iik}^2 \right) r_{iik}^{-\alpha}}{\left(1 + r_{iik}^{-\alpha} \Lambda_i \right)^2} + \tau_{iik}^2 r_{iik}^{-\alpha} \right],$$

and

$$\sum_{l} \left| \mathbf{h}_{\bar{i}ik}^{\mathrm{H}} \tilde{\mathbf{w}}_{\mathrm{a},\bar{i}l} \right|^2 = \sum_{l} \frac{\frac{1}{\zeta_{\mathrm{a},i}} \frac{r_{\bar{i}\bar{i}l}^{-\alpha}}{N} \left(-\frac{\partial \Lambda_i}{\partial \rho} \right)}{\left(1 + r_{\bar{i}\bar{i}l}^{-\alpha} \Lambda_i \right)^2} \left[\frac{\left(1 - \tau_{\bar{i}ik}^2 \right) r_{\bar{i}ik}^{-\alpha}}{\left(1 + r_{\bar{i}ik}^{-\alpha} \Lambda_{\bar{i}} \right)^2} + \tau_{\bar{i}ik}^2 r_{\bar{i}ik}^{-\alpha} \right],$$

respectively. For the power normalization factor $\tilde{\zeta}_{a,i}$ and $\tilde{\zeta}_{a,\bar{i}}$, the deterministic equivalence under large-system regime can be derived in a similar way as

$$\tilde{\zeta}_{a,i} \rightarrow \frac{1}{N} \sum_{l=1}^{K} \frac{r_{iil}^{-\alpha} \left(-\frac{\partial \Lambda_i}{\partial \rho}\right)}{\left(1+r_{iil}^{-\alpha} \Lambda_i\right)^2}, \qquad \tilde{\zeta}_{a,\bar{i}} \rightarrow \frac{1}{N} \sum_{l=1}^{K} \frac{r_{\bar{i}il}^{-\alpha} \left(-\frac{\partial \Lambda_{\bar{i}}}{\partial \rho}\right)}{\left(1+r_{\bar{i}il}^{-\alpha} \Lambda_{\bar{i}}\right)^2}.$$

As such, we have the deterministic equivalence of SIR being as

$$\gamma_{a,ik} = \frac{\left(1 - \tau_{iik}^2\right) \frac{\left(r_{iik}^{-\alpha} \Lambda_i\right)^2}{\left(1+r_{iik}^{-\alpha} \Lambda_i\right)^2} \left|\frac{1}{N} \sum_{l=1}^{K} \frac{r_{iil}^{-\alpha}\left(-\frac{\partial \Lambda_i}{\partial \rho}\right)}{\left(1+r_{iil}^{-\alpha} \Lambda_i\right)^2}\right|^{-1}}{\frac{\left(1-\tau_{iik}^2\right)r_{iik}^{-\alpha}}{\left(1+r_{iik}^{-\alpha} \Lambda_i\right)^2} + \tau_{iik}^2 r_{iik}^{-\alpha} + \frac{\left(1-\tau_{\bar{i}ik}^2\right)r_{\bar{i}ik}^{-\alpha}}{\left(1+r_{\bar{i}ik}^{-\alpha} \Lambda_{\bar{i}}\right)^2} + \tau_{\bar{i}ik}^2 r_{\bar{i}ik}^{-\alpha} + I_a}. \tag{2.51}$$

Finally, by letting $\rho \rightarrow 0$, each term in (2.51) that contains Λ_i, respectively, converges to

$$\frac{\left(r_{iik}^{-\alpha} \Lambda_i\right)^2}{\left(1+r_{iik}^{-\alpha} \Lambda_i\right)^2} = \frac{\left(r_{iik}^{-\alpha} \rho \Lambda_i\right)^2}{\left(\rho + r_{iik}^{-\alpha} \rho \Lambda_i\right)^2} \rightarrow 1, \tag{2.52}$$

$$\frac{\left(1-\tau_{iik}^2\right)r_{iik}^{-\alpha}}{\left(1+r_{iik}^{-\alpha} \Lambda_i\right)^2} = \frac{\left(1-\tau_{iik}^2\right)\rho^2 r_{iik}^{-\alpha}}{\left(\rho + \rho r_{iik}^{-\alpha} \Lambda_i\right)^2} \rightarrow 0, \tag{2.53}$$

$$\left|\frac{1}{N} \sum_{l=1}^{K} \frac{r_{iil}^{-\alpha}\left(-\frac{\partial \Lambda_i}{\partial \rho}\right)}{\left(1+r_{iil}^{-\alpha} \Lambda_i\right)^2}\right|^{-1} \rightarrow \frac{\left(1-\beta-\beta'\right)N}{\sum_{l=1}^{K} r_{iil}^{\alpha}}. \tag{2.54}$$

Lemma 2.2 then follows by substituting (2.52), (2.53), and (2.54) into (2.51).

Proof of Theorem 2.2

We consider a typical UE k of BS i that locates at the origin, and denote the distance between the UE and its associated BS as $r_{iik} = t$ and the distance from the UE to its second closest BS as $r_{\bar{i}ik} = s$. As such, we can approximate the out-of-cell interference I_a by its mean based on Campbell's theorem [30], as follows

$$\mathbb{E}\left[I_a\right] = \mathbb{E}\left[\sum_{x \in \Phi_b \setminus \{i,\bar{i}\}} \frac{g_{xik}}{\|x\|^\alpha}\right]$$

$$= \frac{2\pi \lambda s^{-(\alpha-2)}}{\alpha - 2}. \tag{2.55}$$

By substituting (2.55) into (2.33), the conditional SIR received at the typical UE can be approximated as

$$\gamma_{a,ik} \approx \frac{(1 - \tau_{iik}^2)(1 - \beta - \beta')N}{\left(\tau_{iik}^2 t^{-\alpha} + \tau_{iik}^2 s^{-\alpha} + \frac{2\pi\lambda s^{-(\alpha-2)}}{\alpha-2}\right)(t^\alpha + R_k)}. \tag{2.56}$$

The coverage probability can then be approximated as

$$
\begin{aligned}
\mathbb{P}\left(\gamma_{a,ik} \geq \theta\right) &\approx \mathbb{P}\left(R_k \leq \frac{\left(1 - \tau_{iik}^2\right)\left(1 - \beta - \beta'\right)N\theta^{-1}}{\tau_{iik}^2 t^{-\alpha} + \tau_{iik}^2 s^{-\alpha} + \frac{2\pi\lambda s^{-(\alpha-2)}}{\alpha-2}} - t^\alpha\right) \\
&= \frac{1}{\Gamma(\mu)}\mathbb{E}\left[\gamma\left(\mu, \frac{\mu}{\Omega}\left[\frac{\left(1-\tau^2\right)\left(1-\beta-\beta'\right)Ns^\alpha}{\theta\left(\bar{\tau}^2 + \tau^2\frac{s^\alpha}{t^\alpha} + \frac{2\pi\lambda s^2}{\alpha-2}\right)} - t^\alpha\right]^\eta\right)\right].
\end{aligned}
\tag{2.57}
$$

Furthermore, from the mass transport theorem [30], we have the following relationship:

$$\mathbb{E}\left[\left|C_i^E\right|\right] = 2\mathbb{E}\left[|C_i|\right]. \tag{2.58}$$

where $|\cdot|$ denotes the Lebesgue measure. As a result, the expectation of K' can be calculated as

$$
\begin{aligned}
\mathbb{E}\left[K'\right] &\overset{(a)}{=} \lambda_u\mathbb{E}\left[\left|C_i^E\right|\right] - \lambda_u\mathbb{E}\left[|C_i|\right] \\
&= K,
\end{aligned}
\tag{2.59}
$$

where (a) follows from the fact that mean number of UEs in area A is given by $\lambda_u|A|$. Using this fact, we approximate the random variable K' by its mean K. Theorem 2.2 then follows from (2.57) by deconditioning on t and s, with their pdf given in (2.23) and (2.35), respectively.

Proof of Corollary 2.1

Let $\tau^2 = 0$, $\bar{\tau}^2 = 0$, and $\alpha = 4$. We start with the asymptotic result for the network coverage probability under CEU-ZF. Using the Fubini's theorem [34], coverage probability in (2.32) can be written as

$$\mathbb{P}\left(\gamma_{u,ik} \geq \theta\right) \approx \frac{1}{\Gamma(\mu)}\int_0^\infty \exp\left(-\left(\frac{\Omega}{\mu}t\right)^{\frac{1}{\eta}}\frac{2\pi(\lambda\pi)^2\theta}{(1-\beta)(\alpha-2)N}\right)t^{\mu-1}e^{-t}dt. \tag{2.60}$$

As $N \to \infty$, we have

$$\exp\left(-\left(\frac{\Omega}{\mu}t\right)^{\frac{1}{\eta}}\frac{2\pi(\lambda\pi)^2\theta}{(1-\beta)(\alpha-2)N}\right) \sim 1 - \left(\frac{\Omega}{\mu}t\right)^{\frac{1}{\eta}}\frac{2\pi(\lambda\pi)^2\theta}{(1-\beta)(\alpha-2)N}. \quad (2.61)$$

By substituting (2.61) into (2.60), the coverage probability under CEU-ZF read as

$$\mathbb{P}\left(\gamma_{u,ik} \geq \theta\right) \sim 1 - \frac{\Gamma\left(\frac{1}{\eta}+\mu\right)}{\Gamma(\mu)}\left(\frac{\Omega}{\mu}t\right)^{\frac{1}{\eta}}\frac{(\lambda\pi)^2\theta}{(1-\beta)N}, \qquad \text{as } N \to \infty \quad (2.62)$$

and (2.36) follows from using (2.26) and (2.29) into (2.62).

Next, we deal with the asymptotic result for the network coverage probability under CEA-ZF. When $\tau^2, \bar{\tau}^2 \to 0$, we have the following expression for coverage probability in (2.34) by using the Fubini's theorem [34],

$$\mathbb{P}\left(\gamma_{a,ik} \geq \theta\right)$$
$$\approx \frac{1}{\Gamma(\mu)}\int_0^\infty \left(1+\frac{(\Omega t/\mu)^{\frac{1}{\eta}}(\lambda\pi)^2\theta}{(1-2\beta)N}\right)\exp\left(-\frac{(\Omega t/\mu)^{\frac{1}{\eta}}(\lambda\pi)^2\theta}{(1-2\beta)N}\right)t^{\mu-1}e^{-t}dt$$

$$(2.63)$$

As $N \to \infty$, we have

$$\exp\left(-\frac{(\Omega t/\mu)^{\frac{1}{\eta}}(\lambda\pi)^2\theta}{(1-2\beta)N}\right) \sim 1 - \frac{(\Omega t/\mu)^{\frac{1}{\eta}}(\lambda\pi)^2\theta}{(1-2\beta)N}. \quad (2.64)$$

By substituting (2.64) into (2.63), the coverage probability under CEA-ZF read as

$$\mathbb{P}\left(\gamma_{a,ik} \geq \theta\right) \sim 1 - \frac{\Gamma\left(\frac{2}{\eta}+\mu\right)}{\Gamma(\mu)}\left(\frac{\Omega}{\mu}t\right)^{\frac{2}{\eta}}\frac{(\lambda\pi)^4\theta^2}{(1-2\beta)^2N^2}, \qquad \text{as } N \to \infty \quad (2.65)$$

and (2.37) follows from using (2.26) and (2.30) into (2.65).

References

1. Ericsson, "5G radio access - Capabilities and technologies," *white paper*, Apr. 2016.
2. Nokia Networks, "Ten key rules of 5G deployment - Enabling 1 Tbit/s/km^2 in 2030," *white paper*, Apr. 2015.
3. J. G. Andrews, S. Buzzi, W. Choi, S. V. Hanly, A. Lozano, A. C. Soong, and J. C. Zhang, "What will 5G be?" *IEEE J. Sel. Areas Commun.*, vol. 32, no. 6, pp. 1065–1082, Jun. 2014.
4. T. L. Marzetta, "Noncooperative cellular wireless with unlimited numbers of base station antennas," *IEEE Trans. Wireless Commun.*, vol. 9, no. 11, pp. 3590–3600, Nov. 2010.

5. F. Rusek, D. Persson, B. K. Lau, E. G. Larsson, T. L. Marzetta, O. Edfors, and F. Tufvesson, "Scaling up MIMO: Opportunities and challenges with very large arrays," *IEEE Signal Process. Mag.*, vol. 30, no. 1, pp. 40–60, Oct. 2013.

6. L. Lu, G. Y. Li, A. L. Swindlehurst, A. Ashikhmin, and R. Zhang, "An overview of massive MIMO: Benefits and challenges," *IEEE J. Sel. Topics Signal Process.*, vol. 8, no. 5, pp. 742–758, Oct. 2014.

7. G. Nigam, P. Minero, and M. Haenggi, "Coordinated multipoint joint transmission in heterogeneous networks," *IEEE Trans. Commun.*, vol. 62, no. 11, pp. 4134–4146, Nov. 2014.

8. Z. Xu, C. Yang, G. Y. Li, Y. Liu, and S. Xu, "Energy-efficient CoMP precoding in heterogeneous networks," *IEEE Trans. Signal Process.*, vol. 62, no. 4, pp. 1005–1017, Feb. 2014.

9. R. Tanbourgi, S. Singh, J. G. Andrews, and F. K. Jondral, "A tractable model for noncoherent joint-transmission base station cooperation," *IEEE Trans. Wireless Commun.*, vol. 13, no. 9, pp. 4959–4973, Sep. 2014.

10. A. Lozano, R. W. Heath Jr., and J. G. Andrews, "Fundamental limits of cooperation," *IEEE Trans. Inf. Theory*, vol. 59, no. 9, pp. 5213–5226, Sep. 2013.

11. R. Zakhour and S. V. Hanly, "Base station cooperation on the downlink: Large system analysis," *IEEE Trans. Inf. Theory*, vol. 58, no. 4, pp. 2079–2106, Apr. 2012.

12. R. Bhagavatula and R. W. Heath Jr., "Adaptive limited feedback for sum-rate maximizing beamforming in cooperative multicell systems," *IEEE Trans. Signal Process.*, vol. 59, no. 2, pp. 800–811, Jan. 2011.

13. Y. Huang, S. Durrani, and X. Zhou, "Interference suppression using generalized inverse precoder for downlink heterogeneous networks," *IEEE Wireless Commun. Lett.*, vol. 4, no. 3, pp. 325–328, Jun. 2015.

14. J. Hoydis, K. Hosseini, S. ten Brink, and M. Debbah, "Making smart use of excess antennas: Massive MIMO, small cells, and TDD," *Bell Labs Tech. J.*, vol. 18, no. 2, pp. 5–21, Sep. 2013.

15. E. Björnson, E. G. Larsson, and M. Debbah, "Massive MIMO for maximal spectral efficiency: How many users and pilots should be allocated?" *IEEE Trans. Wireless Commun.*, vol. 15, no. 2, pp. 1293–1308, Feb. 2016.

16. X. Zhu, Z. Wang, C. Qian, L. Dai, J. Chen, S. Chen, and L. Hanzo, "Soft pilot reuse and multicell block diagonalization precoding for massive MIMO systems," *IEEE Trans. Veh. Technol.*, vol. PP, no. 99, pp. 1–1, 2015.

17. J. Hoydis, S. ten Brink, and M. Debbah, "Massive MIMO in the UL/DL of cellular networks: How many antennas do we need?" *IEEE J. Sel. Areas Commun.*, vol. 31, no. 2, pp. 160–171, Feb. 2013.

18. Fujitsu Network Communications, "Enhancing LTE cell-edge performance via PDCCH ICIC," *white paper*, Mar. 2011.

19. E. Björnson and E. G. Larsson, "Three practical aspects of massive MIMO: Intermittent user activity, pilot synchronism, and asymmetric deployment," in *Proc. IEEE Global Telecomm. Conf.*, San Diego, CA, Dec. 2015, pp. 1–6.

20. H. H. Yang, G. Geraci, and T. Q. S. Quek, "Energy-efficient design of MIMO heterogeneous networks with wireless backhaul," *IEEE Trans. Wireless Commun.*, vol. PP, no. 99, pp. 1–1, 2016.

21. S. Singh, X. Zhang, and J. G. Andrews, "Joint rate and SINR coverage analysis for decoupled uplink-downlink biased cell associations in HetNets," *IEEE Trans. Wireless Commun.*, vol. 14, no. 10, pp. 5360–5373, Oct. 2015.

22. D.-T. Lee, "On k-nearest neighbor Voronoi diagrams in the plane," *IEEE Trans. Comput.*, vol. 31, no. 6, pp. 478–487, Jun. 1982.

23. F. Baccelli and A. Giovanidis, "A stochastic geometry framework for analyzing pairwise-cooperative cellular networks," *IEEE Trans. Wireless Commun.*, vol. 14, no. 2, pp. 794–808, Feb. 2015.

24. D. Lopez-Perez, I. Guvenc, and X. Chu, "Mobility management challenges in 3GPP heterogeneous networks," *IEEE Commun. Mag.*, vol. 50, no. 12, pp. 70–78, Dec. 2012.

25. R. Couillet and M. Debbah, *Random matrix methods for wireless communications.* Cambridge University Press, 2011.

26. R. R. Müller, L. Cottatellucci, and M. Vehkapera, "Blind pilot decontamination," *IEEE J. Sel. Topics Signal Process.*, vol. 8, no. 5, pp. 773–786, 2014.

27. T. L. Marzetta and B. M. Hochwald, "Fast transfer of channel state information in wireless systems," *IEEE Trans. Signal Process.*, vol. 54, no. 4, pp. 1268–1278, Apr. 2006.

28. E. Björnson, E. G. Larsson, and T. L. Marzetta, "Massive MIMO: Ten myths and one critical question," *IEEE Commun. Mag.*, vol. 54, no. 2, pp. 114–123, Feb. 2016.

29. A. Kammoun, A. Müller, E. Björnson, and M. Debbah, "Linear precoding based on polynomial expansion: Large-scale multi-cell MIMO systems," *IEEE J. Sel. Topics Signal Process.*, vol. 8, no. 5, pp. 861–875, Jan. 2014.

30. F. Baccelli and B. Blaszczyszyn, *Stochastic Geometry and Wireless Networks. Volumn I: Theory*. Now Publishers, 2009.

31. G. Geraci, H. S. Dhillon, J. G. Andrews, J. Yuan, and I. B. Collings, "Physical layer security in downlink multi-antenna cellular networks," *IEEE Trans. Commun.*, vol. 62, no. 6, pp. 2006–2021, Jun. 2014.

32. R. W. Heath Jr., M. Kountouris, and T. Bai, "Modeling heterogeneous network interference using Poisson point processes," *IEEE Trans. Signal Process.*, vol. 61, no. 16, pp. 4114–4126, Aug. 2013.

33. M. Haenggi, *Stochastic geometry for wireless networks*. Cambridge University Press, 2012.

34. P. Billingsley, *Probability and measure*. John Wiley & Sons, 2008.

35. J. C. S. Santos Filho and M. D. Yacoub, "Simple precise approximations to Weibull sums," *IEEE Commun. Lett.*, vol. 10, no. 8, pp. 614–616, Aug. 2006.

36. H. S. Dhillon, M. Kountouris, and J. G. Andrews, "Downlink MIMO hetnets: Modeling, ordering results and performance analysis," *IEEE Trans. Wireless Commun.*, vol. 12, no. 10, pp. 5208–5222, Oct. 2013.

37. C. Li, J. Zhang, and K. Letaief, "Throughput and energy efficiency analysis of small cell networks with multi-antenna base stations," *IEEE Trans. Wireless Commun.*, vol. 13, no. 5, pp. 2505–2517, May 2014.

38. S. V. Hanly and D. N. C. Tse, "Resource pooling and effective bandwidths in CDMA networks with multiuser receivers and spatial diversity," *IEEE Trans. Inf. Theory*, vol. 47, no. 4, pp. 1328–1351, May 2001.

Chapter 3
Massive MIMO in Small Cell Networks: Wireless Backhaul

Abstract Dense small cell networks are expected to be deployed in the next generation wireless system to provide better coverage and throughput to meet the ever-increasing requirements of high data rate applications. As the trend toward densification calls for more and more wireless links to forward a massive backhaul traffic into the core network, it is critically important to take into account the presence of a wireless backhaul for the energy-efficient design of small cell networks. In this chapter, we develop a general framework to analyze the energy efficiency of a two-tier small cell network with massive MIMO macro base stations and wireless backhaul. Our analysis reveal that under spatial multiplexing, the energy efficiency of a small cell network is sensitive to the network load, and it should be taken into account when controlling the number of users served by each base station. We also demonstrate that a two-tier small cell network with wireless backhaul can be significantly more energy efficient than a one-tier cellular network. However, this requires the bandwidth division between radio access links and wireless backhaul to be optimally designed according to the load conditions.

3.1 Introduction

The next generation of wireless communication systems targets a thousandfold capacity improvement to meet the exponentially growing mobile data demand, and the prospective increase in energy consumption poses urgent environmental and economic challenges [1, 2]. Green communications have become an inevitable necessity, and much effort is being made both in industry and academia to develop new network architectures that can reduce the energy per bit from current levels, thus ensuring the sustainability of future wireless networks [3–7].

3.1.1 Background and Motivation

Since the current growth rate of wireless data exceeds both spectral efficiency advances and availability of new wireless spectrum, a trend toward network densification is essential to respond adequately to the continued surge in mobile data

© The Author(s) 2017
H.H. Yang and T.Q.S. Quek, *Massive MIMO Meets Small Cell*, SpringerBriefs
in Electrical and Computer Engineering, DOI 10.1007/978-3-319-43715-6_3

traffic [8–10]. To this end, small cell networks are proposed to provide higher coverage and throughput by overlaying macro cells with a vast number of low-power small access points, thus offloading traffic and reducing the distance between transmitter and receiver [11, 12]. Forwarding a massive cellular traffic to the backbone network becomes a key problem when small cells are densely deployed, and a wireless backhaul is regarded as the only practical solution where wired links are hardly available [13–17]. However, the power consumption incurred on the wireless backhaul links, together with the power consumed by the multitude of access points deployed, becomes a crucial issue, and an energy-efficient design is necessary to ensure the viability of future small cell networks [18].

Various approaches have been investigated to improve the energy efficiency of small cell networks. Cell size, deployment density, and number of antennas were optimized to minimize the power consumption of small cells [19, 20]. Cognitive sensing and sleep mode strategies were also proposed to turn off inactive access points and enhance the energy efficiency [21, 22]. A further energy efficiency gain was shown to be attainable by serving users that experience better channel conditions, and by dynamically assigning users to different tiers of the network [23, 24]. Although various studies have been conducted on the energy efficiency of small cell networks, the impact of a wireless backhaul has typically been neglected. On the other hand, the power consumption of backhauling operations at small cell access points (SAPs) might be comparable to the amount of power necessary to operate macrocell base stations (MBSs) [25–27]. Moreover, since it is responsible to aggregate traffic from SAPs toward MBSs, the backhaul may significantly affect the rates and therefore the energy efficiency of the entire network. With a potential evolution toward dense infrastructures, where many small access points are expected to be used, it is of critical importance to take into account the presence of a wireless backhaul for the energy-efficient design of heterogeneous networks.

3.1.2   Approach and Main Outcomes

Our main goal in this chapter is to study the energy-efficient design of small cell networks with wireless backhaul. In particular, we consider a two-tier small cell network which consists of MBSs and SAPs, where SAPs are connected to MBSs via a multiple-input-multiple-output (MIMO) wireless backhaul that uses a fraction of the total available bandwidth. We undertake an analytical approach to derive data rates and power consumption for the entire network in the presence of both uplink (UL) and downlink (DL) transmissions and spatial multiplexing, which is a practical scenario that has not yet been addressed. Similar to the framework we developed in Chap. 2, we model the spatial locations of MBSs, SAPs, and UEs as independent homogeneous Poisson point processes (PPPs), and analyze the energy efficiency by combining tools from stochastic geometry and random matrix theory. The analysis enable us to take a complete treatment of all the key features in a small cell network, i.e., interference, load, deployment strategy, and capability of the wireless infrastructure components. With the developed framework, we can explicitly characterize the power consumption of the small cell network due to signal processing operations in macro cells, small cells, and wireless backhaul, as well as the rates and ultimately the energy efficiency of the whole network. The main contributions in this chapter are summarized below.

- We provide a general toolset to analyze the energy efficiency of a two-tier small -network with wireless backhaul. Our model accounts for both UL and DL transmissions and spatial multiplexing, for the bandwidth and power allocated between macro cells, small cells, and backhaul, and for the infrastructure deployment strategy.
- We combine tools from stochastic geometry and random matrix theory to derive the uplink and downlink rates of macro cells, small cells, and wireless backhaul. The resulting analysis is tractable and captures the effects of multiantenna transmission, fading, shadowing, and random network topology.
- Using the developed framework, we find that the energy efficiency of a small cell network is sensitive to the load conditions of the network, thus establishing the importance of scheduling the right number of UEs per base station. Moreover, by comparing the energy efficiency under different deployment scenarios, we find that such property does not depend on the infrastructure.
- We show that if the wireless backhaul is not allocated sufficient resources, then the energy efficiency of a two-tier small cell network with wireless backhaul can be worse than that of a one-tier cellular network. However, the two-tier small cell network can achieve a significant energy efficiency gain if the backhaul bandwidth is optimally allocated according to the load conditions of the network.

The remainder of this chapter is organized as follows. The system model is introduced in Sect. 3.2. In Sect. 3.3, we detail the power consumption of a heterogeneous network with wireless backhaul. In Sect. 3.4, we analyze the data rates and the energy efficiency, and we provide simulations that confirm the accuracy of our analysis. Numerical results are shown in Sect. 3.5 to give insights into the energy-efficient design of a HetNet with wireless backhaul. The chapter is concluded in Sect. 3.6.

3.2 System Model

3.2.1 Topology and Channel

We study a two-tier small cell network which consists of MBSs, SAPs, and UEs, as depicted in Fig. 3.1. The spatial locations of MBSs, SAPs, and UEs follow independent PPPs Φ_m, Φ_s, and Φ_u, with spatial densities λ_m, λ_s, and λ_u, respectively. All MBSs, SAPs, and UEs are equipped with M_m, M_s, and 1 antennas, respectively, each UE associates with the base station that provides the largest average received power, and each SAP associates with the closest MBS. The links between MBSs and UEs,

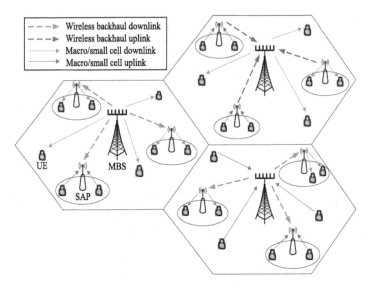

Fig. 3.1 Illustration of a two-tier small cell network with wireless backhaul

SAPs and UEs, and MBSs and SAPs are referred to as *macro cell links*, *small cell links*, and *backhaul links*, respectively. In light of its higher spectral efficiency [28], we consider spatial multiplexing where each MBS and each SAP simultaneously serve K_m and K_s UEs, respectively. In practice, due to a finite number of antennas, MBSs and SAPs use traffic scheduling to limit the number of UEs served to $K_m \leq M_m$ and $K_s \leq M_s$ [29]. Similarly, each MBS limits to K_b the number of SAPs served on the backhaul, with $K_b M_s \leq M_m$. The MIMO dimensionality ratio for linear processing on macrocells, small cells, and backhaul is denoted by $\beta_m = \frac{K_m}{M_m}$, $\beta_s = \frac{K_s}{M_s}$, and $\beta_b = \frac{K_b M_s}{M_m}$, respectively.

In this work, we consider a co-channel deployment of small cells with the macro cell tier, i.e., macro cells and small cells share the same frequency band for transmission.[1] In order to avoid severe interference which may degrade the performance of the network, we assume that the access and backhaul links share the same pool of radio resources through orthogonal division, i.e., the total available bandwidth is divided into two portions, where a fraction ζ_b is used for the wireless backhaul, and the remaining $(1 - \zeta_b)$ is shared by the radio access links (macro cells and small cells) [13, 15, 34, 35]. In order to adapt the radio resources to the variation of the DL/UL traffic demand, we assume that MBSs and SAPs operate in a dynamic time division duplex (TDD) mode [36, 37], where at every time slot, all MBSs and

---

[1]Many frequency planning possibilities exist for MBSs and SAPs, where the optimal solution is traffic load dependent. Though a non-co-channel allocation is justified for highly dense scenarios [30–32], in some cases a co-channel deployment may be preferred from an operator's perspective, since MBSs and SAPs can share the same spectrum thus improving the spectral utilization ratio [33].

SAPs independently transmit in downlink with probabilities τ_m, τ_s, and τ_b on the macro cell, small cell, and backhaul, respectively, and they transmit in uplink for the remaining time.[2] We model the channels between any pair of antennas in the network as independent, narrowband, and affected by three attenuation components, namely, small-scale Rayleigh fading, shadowing S_D and S_B for data link and backhaul link, respectively, and large-scale path loss, where α is the path loss exponent and the shadowing satisfies $\mathbb{E}[S_D^{\frac{2}{\alpha}}] < \infty$ and $\mathbb{E}[S_B^{\frac{2}{\alpha}}] < \infty$, and by thermal noise with variance σ^2. We finally assume that all MBSs and SAPs use a zero forcing (ZF) scheme for both transmission and reception, due to its practical simplicity [38].[3]

3.2.2 Energy Efficiency

We consider the power consumption due to transmission and signal processing operations performed on the entire network, therefore energy-efficiency tradeoffs will be such that savings at the MBSs and SAPs are not counteracted by increased consumption at the UEs, and vice versa [4, 43]. To this end, we identify the three aspects as the major power consumption in the network, namely, the power spent on macro cells, small cells, and wireless backhaul. Consistent with previous work [43–46], we account for the power consumption due to transmission, encoding, decoding, and analog circuits.

Let $\mathscr{P}[\frac{W}{m^2}]$ be the total power consumption per area, which includes the power consumed on all links. We denote by $\mathscr{R}[\frac{bit}{m^2}]$ the sum rate per unit area of the network, i.e., the total number of bits per second successfully transmitted per square meter. The energy efficiency $\eta = \frac{\mathscr{R}}{\mathscr{P}}$ is then defined as the number of bits successfully transmitted per joule of energy spent [43, 47]. For the sake of clarity, the main notations used in this paper are summarized in Table 3.1.

3.3 Power Consumption

In this section, we model in detail the power consumption of the small cell network with wireless backhaul.

To start with, notice that each UE associates with the base station, i.e., MBS or SAP that provides the largest average received power, the probability that a UE associates to a MBS or to a SAP can be respectively calculated as [48]

---

[2]We note that different SAPs and MBSs may have different uplink/downlink resource partitions for their associated UEs. Since the aggregate interference is affected by the average value of such partitions, we assume fixed and uniform uplink/downlink partitions.

[3]Note that the results involving the machinery of random matrix theory can be adjusted to account for different transmit precoders and receive filters, imperfect channel state information, and antenna correlation [39–42].

Table 3.1 Notation summary

Notation	Definition
$\mathscr{P}; \mathscr{R}; \eta$	Power per area; rate per area; energy efficiency
$R_{\rm m}^{\rm DL}; R_{\rm s}^{\rm DL}; R_{\rm b}^{\rm DL}$	Downlink rate on macrocells, small cells, and backhaul
$R_{\rm m}^{\rm UL}; R_{\rm s}^{\rm UL}; R_{\rm b}^{\rm UL}$	Uplink rate on macrocells, small cells, and backhaul
$P_{\rm mt}; P_{\rm st}; P_{\rm ut}$	Transmit power for MBSs, SAPs, and UEs
$P_{\rm mb}; P_{\rm sb}$	Backhaul transmit power for MBSs and SAPs
$P_{\rm mc}; P_{\rm sc}$	Analog circuit power consumption at macrocells and small cells
$P_{\rm me}; P_{\rm se}; P_{\rm ue}$	Encoding power per bit on macrocells, small cells, and backhaul
$P_{\rm md}; P_{\rm sd}; P_{\rm ud}$	Decoding power per bit on macrocells, small cells, and backhaul
$\Phi_{\rm m}; \Phi_{\rm s}; \Phi_{\rm u}$	PPPs modeling locations of MBSs, SAPs, and UEs
$\lambda_{\rm m}; \lambda_{\rm s}; \lambda_{\rm u}$	Spatial densities of MBSs, SAPs, and UEs
$A_{\rm m}; A_{\rm s}$	Association probabilities for MBSs and SAPs
$M_{\rm m}; M_{\rm s}$	Number of transmit antennas per MBSs and SAPs
$K_{\rm m}; K_{\rm s}; K_{\rm b}$	UEs served per macrocell and small cell; SAPs per MBSs on backhaul
$\tau_{\rm m}; \tau_{\rm s}; \tau_{\rm b}$	Fraction of time in DL for macrocells, small cells, and backhaul
$\zeta_{\rm b}; \alpha$	Fraction of bandwidth for backhaul; path loss exponent
$S_{\rm D}; S_{\rm B}$	Shadowing on radio access link and wireless backhaul

$$A_{\rm m} = \frac{\lambda_{\rm m} P_{\rm mt}^{\frac{2}{\alpha}}}{\lambda_{\rm m} P_{\rm mt}^{\frac{2}{\alpha}} + \lambda_{\rm s} P_{\rm st}^{\frac{2}{\alpha}}} \tag{3.1}$$

and

$$A_{\rm s} = \frac{\lambda_{\rm s} P_{\rm st}^{\frac{2}{\alpha}}}{\lambda_{\rm m} P_{\rm mt}^{\frac{2}{\alpha}} + \lambda_{\rm s} P_{\rm st}^{\frac{2}{\alpha}}}. \tag{3.2}$$

In the remainder of this chapter, we make the assumption that the number of UEs, the number of SAPs associated to a MBS, and the number of UEs associated to a SAP by constant values $K_{\rm m}$, $K_{\rm b}$, and $K_{\rm s}$, respectively, which are upper bounds imposed by practical antenna limitations at MBSs and SAPs.[4]

The assumption above is motivated by the fact that the number of UEs $N_{\rm m}$ served by a MBS has distribution [48]

$$\mathbb{P}(N_{\rm m} = n) = \frac{3.5^{3.5} \Gamma(n + 3.5) \left(\frac{\lambda_{\rm m}}{A_{\rm m} \lambda_{\rm u}}\right)^{3.5}}{\Gamma(3.5) n! \left(1 + 3.5 \lambda_{\rm m}/\lambda_{\rm u}\right)^{n+3.5}}, \tag{3.3}$$

---

[4]The number of base station antennas imposes a constraint on the maximum number of UEs scheduled for transmission. In fact, under linear precoding, the number of scheduled UEs should not exceed the number of antennas, in order for the achievable rate not to be significantly degraded [49–51].

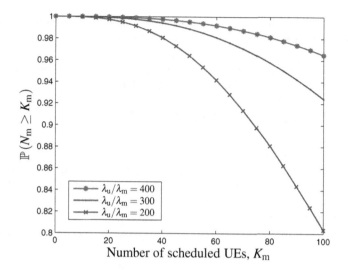

Fig. 3.2 Complementary cumulative distribution function (CCDF) of the number of UEs N_m associated to a MBS, where K_m is the maximum number of UEs that can be served due to antenna limitations

where $\Gamma(\cdot)$ is the gamma function. Let K_m be a limit on the number of users that can be served by a MBS, the probability that a MBS serves less than K_m UEs is given by

$$
\begin{aligned}
\mathbb{P}(N_m < K_m) &= \sum_{n=0}^{K_m-1} \frac{3.5^{3.5}\,\Gamma(n+3.5)\left(\frac{\lambda_m}{A_m\lambda_u}\right)^{3.5}}{\Gamma(3.5)n!\,(1+3.5\lambda_m/\lambda_u)^{n+3.5}} \\
&\leq \left(\frac{2\lambda_m}{\lambda_u}\right)^{3.5} \sum_{n=0}^{K_m-1} \frac{\Gamma(n+3.5)}{n!}\,\frac{3.5^{3.5}}{\Gamma(3.5)},
\end{aligned}
\tag{3.4}
$$

which rapidly tends to zero as $\frac{\lambda_u}{\lambda_m}$ grows. This indicates that in a practical network with a high density of UEs, i.e., where $\lambda_u \gg \lambda_m$, each MBS serves K_m UEs with probability almost one. Figure 3.2 shows the probability $\mathbb{P}(N_m \geq K_m)$ that a MBS has at least K_m UEs to serve, where values of $\mathbb{P}(N_m \geq K_m)$ are plotted for three UE–MBS density ratios λ_u/λ_m, and for various numbers of scheduled users K_m. It can be seen that $\mathbb{P}(N_m \geq K_m) \approx 1$ for moderate-to-high UE densities and low-to-moderate values of K_m, therefore confirming that each MBS tends to serve a fixed number K_m of UEs with probability one. A similar approach can be used to show that $\mathbb{P}(N_s < K_s) \approx 0$ and $\mathbb{P}(N_b < K_b) \approx 0$ when $\lambda_u \gg \lambda_m$ and $\lambda_s \gg \lambda_m$, respectively, and therefore each SAP serves K_s UEs and each MBS serves K_b SAPs on the backhaul with probability almost one.

In the following, we use the power consumption model introduced in [43], which captures all the key contributions to the power consumption of signal processing operations. This model is flexible since the various power consumption values can be tuned according to different scenarios. We note that the results presented in this paper hold under more general conditions and apply to different power consumption models [52, 53].

Under the previous assumption, and by using the model in [43], we can write the power consumption on each macro cell link as follows

$$P_m = \tau_m P_{mt} + (1 - \tau_m)\, K_m P_{ut} + \tau_m K_m\, (P_{me} + P_{ud})\, R_m^{DL}$$
$$+ P_{mc} + (1 - \tau_m)\, K_m\, (P_{md} + P_{ue})\, R_m^{UL}, \qquad (3.5)$$

where P_{mt} and P_{ut} are the DL and UL transmit power from the MBS and the K_m UEs, respectively, P_{mc} is the analog circuit power consumption, P_{me} and P_{md} are encoding and decoding power per bit of information for MBS, while P_{ue} and P_{ud} are encoding and decoding power per bit of information for UE, and R_m^{DL} and R_m^{UL} denote the DL and UL rates for each MBS–UE pair. The analog circuit power can be modeled as [43]

$$P_{mc} = P_{mf} + P_{ma} M_m + P_{ua} K_m, \qquad (3.6)$$

where P_{mf} is a fixed power accounting for control signals, baseband processor, local oscillator at MBS, cooling system, etc., P_{ma} is the power required to run each circuit component attached to the MBS antennas, such as converter, mixer, and filters, P_{ua} is the power consumed by circuits to run a single-antenna UE. Under this model, the total power consumption on the macrocell can be written as

$$P_m = \tau_m P_{mt} + (1 - \tau_m)\, K_m P_{ut} + \tau_m K_m (P_{me} + P_{ud}) R_m^{DL}$$
$$+ P_{mf} + P_{ma} M_m + P_{ua} K_m + (1 - \tau_m) K_m (P_{md} + P_{ue}) R_m^{UL}. \qquad (3.7)$$

Through a similar approach, the power consumption on each small cell and backhaul link can be written as

$$P_s = \tau_s P_{st} + (1 - \tau_s)\, K_s P_{ut} + P_{sf} + \tau_s K_s\, (P_{se} + P_{ud})\, R_s^{DL}$$
$$+ P_{sa} M_s + P_{ua} K_s + (1 - \tau_s)\, K_s\, (P_{sd} + P_{ue})\, R_s^{UL} \qquad (3.8)$$

and

$$P_b = \tau_b P_{mb} + (1 - \tau_b)\, K_b P_{sb} + \tau_b K_b K_s\, (P_{me} + P_{sd})\, R_b^{DL}$$
$$+ P_{ma} M_m + K_b M_s P_{sa} + (1 - \tau_b)\, K_b K_s\, (P_{md} + P_{se})\, R_b^{UL}, \qquad (3.9)$$

respectively, the analog circuit power consumption in (3.9) accounts for power spent on out of band SAPs. In the above equations, P_{st} is the transmit power on a small cell, P_{mb} and P_{sb} are the powers transmitted by MBSs and SAPs on the backhaul, and P_{sf} and P_{sa} are the small-cell equivalents of P_{mf} and P_{ma}. Moreover, R_s^{DL} and

R_s^{UL} denote the DL and UL rates for each SAP–UE pair, and R_b^{DL} and R_b^{UL} denote the DL and UL rates for each wireless backhaul link.

With the above results, the average power consumption per area can be expressed as

$$\mathcal{P} = P_m \lambda_m + P_s \lambda_s + P_b \lambda_m, \tag{3.10}$$

where P_m, P_s, and P_b are given, respectively, in (3.7), (3.8), and (3.9).

3.4 Rates and Energy Efficiency

In this section, we analyze the data rates and the energy efficiency of a small cell network with wireless backhaul. Particularly, we combine tools from stochastic geometry and random matrix theory to derive the uplink and downlink rates of macro cells, small cells, and wireless backhaul. The analytical expressions provided in this section are tight approximations of the actual data rates. For a better readability, proofs and mathematical derivations have been relegated to the Appendix.

To start with, we consider a typical DL transmission link between a typical UE located at the origin and served by its associated MBS. Note that under dynamic TDD [36, 37], the DL communication is corrupted by DL interference from other MBSs and SAPs, and by UL interference from UEs that associated with other MBSs and SAPs. Results from stochastic geometry indicates that the UL interference from UEs that associated with MBSs follow a homogeneous PPP with density $(1 - \tau_m)\lambda_m K_m$, and similarly, and similarly, the UL interference from UEs that associated with SAPs follow a homogeneous PPP with density $(1 - \tau_s)\lambda_s K_s$. The UL interference from UEs that associated with SAPs follow a homogeneous PPP with density $(1 - \tau_s)\lambda_s K_s$. Using composition theorem [54], we have the UL interfering UEs follow a PPP with density $\tilde{\lambda}_u = (1 - \tau_m)\lambda_m K_m + (1 - \tau_s)\lambda_s K_s$.

The large antenna array at MBS allows us to apply random matrix theory tools to obtain the DL rate on a macro cell link.

Lemma 3.1 *The downlink rate on a macrocell is given by*

$$R_m^{DL} = (1 - \zeta_b) \int_0^\infty \int_0^\infty \frac{e^{-\sigma^2 z}}{z \ln 2} \left(1 - e^{-z v_m^D}\right) \exp\left(-\frac{2\pi^2 \tilde{\lambda}_u P_{ut}^\delta \mathbb{E}[S_D^\delta] z^\delta}{\alpha \sin\left(\frac{2\pi}{\alpha}\right)}\right)$$

$$\times \exp\left(-\tau_m a_m \mathscr{C}_{\alpha, K_m}(z P_{mt}, t)\left(\frac{z P_{mt}}{K_m}\right)^\delta - \tau_s a_s \mathscr{C}_{\alpha, K_s}(z P_{mt}, t)\left(\frac{z P_{st}}{K_s}\right)^\delta\right) f_{L_m}(t) dt dz, \tag{3.11}$$

where $\delta = 2/\alpha$, $a_m = \lambda_m \pi \mathbb{E}[S_D^\delta]$, $a_s = \lambda_s \pi \mathbb{E}[S_D^\delta]$, $\tilde{\lambda}_u = (1 - \tau_m)\lambda_m K_m + (1 - \tau_s)\lambda_s K_s$, while v_m^D, $f_{L_m}(t)$, and $\mathscr{C}_{\alpha, K}(z, t)$ given as follows

$$v_m^D = \frac{P_{mt}\,(1 - \beta_m)\,(G_m)^{\frac{\alpha}{2}}}{\beta_m \Gamma\left(1 + \frac{\alpha}{2}\right)}, \tag{3.12}$$

$$f_{L_m}(t) = G_m \delta x^{\delta - 1} \exp\left(-G_m x^\delta\right), \quad x \geq 0 \tag{3.13}$$

$$\mathcal{C}_{\alpha, K}(z, t) = \frac{2}{\alpha} \sum_{n=1}^{K} \binom{K}{n} \left[B\left(1; K - n + \frac{2}{\alpha}, n - \frac{2}{\alpha}\right) \right.$$
$$\left. - B\left(\left(1 + \frac{s}{tK}\right)^{-1}; K - n + \frac{2}{\alpha}, n - \frac{2}{\alpha}\right) \right] \tag{3.14}$$

with $G_m = a_m + a_s\,(P_{st}/P_{mt})^\delta$, and $B(x; y, z) = \int_0^x t^{y-1}(1-t)^{z-1}dt$ the incomplete Beta function.

The proof of this lemma is given in Appendix section "Proof of Lemma 3.1". In Fig. 3.3, we provide a comparison between the simulated macrocell downlink rate and the analytical result obtained in Lemma 3.1 with different antenna numbers at the MBS. The downlink rate is plotted versus the transmit power at the MBSs. It can be seen that analytical results and simulations fairly well match, thus confirming the accuracy of Lemma 3.1.

We next deal with the analysis to the uplink achievable rate of an MBS UE. Note that in the downlink, due to the maximum received power association, interfering base station cannot be located closer to the typical user than the tagged base station, i.e., an exclusion region exists where the distance between a UE and the interfering base stations is bounded away from zero. However in the uplink, since PPP deployment assumption ignores a minimum inter-site distance between base stations, it can happen that an interfering base station locates arbitrarily close to a typical MBS, i.e.,

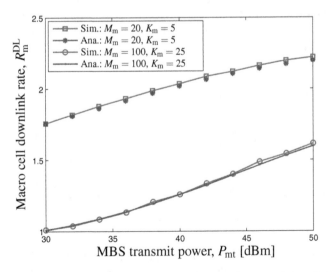

Fig. 3.3 Comparison of the simulations and numerical results for macrocell downlink rate

the distance between a MBS and the interfering base stations can be arbitrarily small. In the following, we treat the latter as a composition of three independent PPPs with different spatial densities. We then apply stochastic geometry to obtain the macrocell uplink rate as follows:

Lemma 3.2 *The uplink rate on a macro cell is given by*

$$
R_m^{UL} = (1 - \zeta_b) \int_0^\infty \int_0^\infty \frac{\left(1 - e^{-zv_m^U/t}\right)}{ze^{\sigma^2 z} \ln 2} \exp\left\{-\tilde{\lambda}_u \pi \mathbb{E}[S_D^\delta] \int_0^\infty \frac{1 - e^{-G_m u}}{1 + z^{-1} u^{\frac{1}{\delta}}/P_{ut}} du \right.
$$
$$
\left. - \frac{\Gamma(1+\delta) \, \delta \pi^2 z^\delta}{\sin(\delta\pi)} \left[\frac{\tau_m a_m P_{mt}^\delta \prod_{i=1}^{K_m-1}(i+\delta)}{\Gamma(K_m) K_m^\delta} + \frac{\tau_s a_s P_{st}^\delta \prod_{i=1}^{K_s-1}(i+\delta)}{\Gamma(K_s) K_s^\delta} \right] \right\} f_{L_m}(t) dt dz
$$

$$(3.15)$$

with $v_m^U = (1 - \beta_m) M_m P_{mt}$.

The proof is given in Appendix section "Proof of Lemma 3.2."

In order to derive the downlink and uplink rate of an SAP UE, we apply similar trick as we used in the derivation of macrocell rate. However, unlike the macrocell, due to the relatively small number of antennas at the SAPs, random matrix theory tools cannot be employed to calculate the rate on a small cell. We therefore use the effective channel distribution as follows:

Lemma 3.3 *The downlink rate on a small cell is given by*

$$
R_s^{DL} = \int_0^\infty \int_0^\infty \frac{(1 - \zeta_b)}{ze^{\sigma^2 z} \ln 2} \left(1 - \frac{1}{(1 + zP_{st}t^{-1}/K_s)^{\Delta_s}}\right) \exp\left(-\frac{2\pi^2 \tilde{\lambda}_u P_{ut}^{\frac{2}{\alpha}} \mathbb{E}[S_D^{\frac{2}{\alpha}}] z^{\frac{2}{\alpha}}}{\alpha \sin\left(\frac{2\pi}{\alpha}\right)}\right)
$$
$$
\times \exp\left(-\tau_s a_s \mathscr{C}_{\alpha, K_s}(zP_{st}, t) \left(\frac{zP_{st}}{K_s}\right)^\delta - \tau_m a_m \mathscr{C}_{\alpha, K_m}(zP_{st}, t) \left(\frac{zP_{mt}}{K_m}\right)^\delta\right) f_{L_s}(t) dt dz,
$$

$$(3.16)$$

where $\Delta_s = M_s - K_s + 1$, *and* $f_{L_s}(t)$ *is given as*

$$
f_{L_s}(t) = G_s \delta t^{\delta-1} \exp\left(-G_s t^\delta\right), \quad t \geq 0
$$

$$(3.17)$$

with $G_s = a_s + a_m (P_{mt}/P_{st})^\delta$.

Following a similar approach as the one in Lemma 3.2, we can obtain the uplink rate on a small cell.

Lemma 3.4 *The uplink rate on a small cell is given by*

$$
R_s^{UL} = \int_0^\infty \frac{(1 - \zeta_b)}{ze^{\sigma^2 z} \ln 2} \left[1 - \int_0^\infty \frac{f_{L_s}(t) dt}{(1 + zP_{ut}/t)^{\Delta_s}}\right] \exp\left\{-\tilde{\lambda}_u \pi \mathbb{E}[S_D^\delta] \int_0^\infty \frac{1 - e^{-G_s z}}{1 + z^{-1} u^{\frac{1}{\delta}}/P_{ut}} du \right.
$$
$$
\left. - \frac{\Gamma(1+\delta) \, \delta \pi^2 z^\delta}{\sin(\delta\pi)} \left[\frac{\tau_s a_s P_{st}^\delta \prod_{i=1}^{K_s-1}(i+\delta)}{\Gamma(K_s) K_s^\delta} + \frac{\tau_m a_m P_{mt}^\delta \prod_{i=1}^{K_m-1}(i+\delta)}{\Gamma(K_m) K_m^\delta} \right] \right\} dz.
$$

$$(3.18)$$

The proof of Lemmas 3.3 and 3.4 can be found in [55].

Now, it remains to derive the downlink and uplink rates on the wireless backhaul. In the communication between MBS and SAP, each end of the transmission link involves multiple antennas. For this scenario, it has been shown that using block diagonalization (BD) is the optimal way to achieve capacity. However, there are no closed form expression is available for the rate achievable by BD. To this end, we treat each antenna of SAPs as an individual UE, and use ZF at the MBS to do the precoding/decoding. Although ZF is suboptimal compared to the BD, we will show by Fig. 3.4 that the rate gap between these two transmission schemes is limited, and that the rates under BD and ZF follow a similar trend. Therefore, our findings on the energy efficiency tradeoffs remain valid irrespective of the scheme used. In the following, we present the uplink and downlink rate of the wireless backhaul, and then show the simulation comparison to confirm the above claim.

Lemma 3.5 *The downlink rate on the wireless backhaul is given by*

$$
R_b^{DL} = \frac{\zeta_b M_s}{K_s} \int_0^\infty \int_0^\infty \frac{\left(1 - e^{-z v_b^D}\right)}{z e^{\sigma^2 z} \ln 2} \exp\left(-\tau_b a_b \mathscr{C}_{\alpha, K_b M_s}(z P_{mb}, t) \left(\frac{z P_{mb}}{K_b M_s}\right)^\delta\right)
$$

$$
\times \exp\left(-\frac{\Gamma(1+\delta)\, \delta\pi^2 z^\delta P_{sb}^\delta}{\sin(\delta\pi)\Gamma(M_s) M_s^\delta} \mathbb{E}[S_B^\delta](1-\tau_b)\lambda_s \prod_{i=1}^{M_s-1}(i+\delta)\right) f_{L_b}(t)\,dt\,dz,
$$

$$
\tag{3.19}
$$

where $a_b = \lambda_m \pi \mathbb{E}[S_B^\delta]$, $f_{L_b}(t)$ and v_b^D are given as

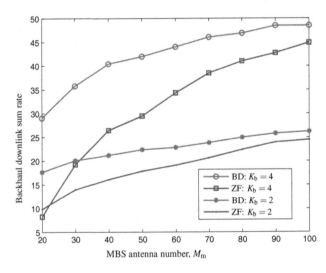

Fig. 3.4 Comparison of the simulations and numerical results for macrocell downlink rate

$$f_{L_b}(t) = a_b \delta t^{\delta-1} \exp(-a_b t^\delta), \; t > 0 \tag{3.20}$$

$$v_b^D = \frac{P_{mb}(1 - \beta_b) a_b^\delta}{\beta_b \Gamma(1 + 1/\delta)}. \tag{3.21}$$

Lemma 3.6 *The uplink rate on the wireless backhaul is given by*

$$R_b^{UL} = \frac{\zeta_b M_s}{K_s} \int_0^\infty \int_0^\infty \frac{\left(1 - e^{-zv_b^U/t}\right)}{z e^{\sigma^2 z} \ln 2} \exp\left\{ -\frac{\tau_b a_b \Gamma(1+\delta) \delta \pi^2 P_{mb}^\delta z^\delta \prod_{i=1}^{K_b M_s - 1}(i + \delta)}{\sin(\delta\pi)(M_s K_b)^\delta \Gamma(M_s K_b)} \right\}$$

$$\times \exp\left\{ -(1-\tau_b) a_b K_b \sum_{n=1}^{M_s} \binom{M_s}{n} \int_0^\infty \frac{\left(zu^{-1/\delta} P_{sb}/M_s\right)^n \left(1 - e^{-a_b u}\right)}{\left(1 + zu^{-1/\delta} P_{sb}/M_s\right)^{M_s}} du \right\} f_{L_b}(t) dt dz \tag{3.22}$$

where $v_b^U = (1 - \beta_b) M_m P_{sb}$.

In Fig. 3.4 compares the rate on the wireless backhaul under BD and ZF, respectively, with different numbers of SAPs. It can be seen that although ZF achieves a lower rate than BD, the rate gap is limited as the antenna number grows, and the rates under BD and ZF follow a similar trend. Therefore, the conclusions drawn in this paper on the energy efficiency tradeoffs remain valid irrespective of the scheme used.

We can now write the data rate per area in a small cell network with wireless backhaul by combining results from above.

Lemma 3.7 *The sum rate per area in a small cell network with wireless backhaul is given by*

$$\mathcal{R} = B\left(K_m \lambda_m + K_s \lambda_s\right) \left\{ A_m\left[\tau_m R_m^{DL} + (1 - \tau_m) R_m^{UL}\right]\right.$$

$$\left. + A_s\left[\tau_s \min\left\{R_s^{DL}, R_b^{DL}\right\} + (1 - \tau_s) \min\left\{R_s^{UL}, R_b^{UL}\right\}\right]\right\}, \tag{3.23}$$

where B is the total available bandwidth, and $R_m^{DL}, R_m^{UL}, R_s^{DL}, R_s^{UL}, R_b^{DL}$, *and* R_b^{UL} *are given in* (3.11), (3.15), (3.16), (3.18), (3.19), *and* (3.22), *respectively.*

Proof See Appendix "Proof of Lemma 3.7".

Note that the energy efficiency is obtained as the ratio between the data rate per area and the power consumption per area. We finally obtain the energy efficiency of a heterogeneous network with wireless backhaul, defined as the number of bits successfully transmitted per joule of energy spent.

Theorem 3.1 *The energy efficiency* η *of a heterogeneous network with wireless backhaul is given by*

$$\eta = \frac{B \, (K_m \lambda_m + K_s \lambda_s)}{P_m \lambda_m + P_s \lambda_s + P_b \lambda_m} \left(A_m \left[\tau_m R_m^{DL} + (1 - \tau_m) R_m^{UL} \right] \right.$$
$$\left. + A_s \left[\tau_s \min \left\{ R_s^{DL}, R_b^{DL} \right\} + (1 - \tau_s) \min \left\{ R_s^{UL}, R_b^{UL} \right\} \right] \right). \tag{3.24}$$

Equation (3.24) quantifies how all the key features of a small cell network, i.e., interference, deployment strategy, and capability of the wireless infrastructure components, affect the energy efficiency when a wireless backhaul is used to forward traffic into the core network. Several numerical results based on (3.24) will be shown in Sect. 3.5 to give more practical insights.

3.5 Numerical Results

In this section, we provide numerical results to show how the energy efficiency is affected by various network parameters and to give insights into the optimal design of a small cell network with wireless backhaul. As an example, we consider two different deployment scenarios, namely, (i) *femto cells* that consist of a dense deployment of low-power SAPs with a small number of antennas, and (ii) *pico cells* that have a less dense deployment of larger and more powerful SAPs. We refer to *light load* and *heavy load* conditions as the ones of a network with $\beta_m = \beta_s = \beta_b = 0.25$ and $0.9 \leq \beta_m, \beta_s, \beta_b < 1$, respectively. The network is considered to be operating at 2 GHz, with path loss exponent set to $\alpha = 3.8$ to model an urban scenario, the shadowing S_B and S_D are set to be lognormal distributed as $S_B = 10^{\frac{X_B}{10}}$ and $S_D = 10^{\frac{X_D}{10}}$, where $X_B \sim N(0, \sigma_B^2)$ and $X_D \sim N(0, \sigma_D^2)$, with $\sigma_B = 3$ dB and $\sigma_D = 6$ dB, respectively [56]. In addition, we equal the backhaul transmit power to the radio access power, i.e., $P_{mb} = P_{mt}$, $P_{sb} = P_{st}$. All other system and power consumption parameters are set as follows: $P_{mt} = 47.8$ dBm, for pico cell SAPs $P_{st} = 30$ dBm, for femto cell SAP $P_{st} = 23.7$ dBm, $P_{ut} = 17$ dBm, $P_{ma} = 1$ W, for pico cell SAPs $P_{sa} = 0.8$ W, for femto cell SAP $P_{sa} = 0.8$ W, $P_{ua} = 0.1$ W [32]; $P_{mf} = 225$ W, for pico cell SAPs $P_{sf} = 7.3$ W, for femto cell SAPs $P_{sf} = 5.2$ W [52, 53]; $P_{me} = 0.1$ W/Gb, $P_{md} = 0.8$W/Gb, $P_{se} = 0.2$ W/Gb, $P_{sd} = 1.6$ W/Gb, $P_{ue} = 0.3$ W/Gb, $P_{md} = 2.4$ W/Gb [43].

Results from Fig. 3.5 illustrate the effect of network load on the energy efficiency. Particularly, we compare the energy efficiency of small cell networks that use pico cells and femto cells in Fig. 3.5a, under various load conditions and for different portions of the bandwidth allocated to the wireless backhaul. The figure shows that femto cell and pico cell deployments exhibit similar performance in terms of energy efficiency. Moreover, Fig. 3.5a shows that the energy efficiency of the network is highly sensitive to the portion of bandwidth allocated to the backhaul, and that there is an optimal value of ζ_b which maximizes the energy efficiency. This optimal value of ζ_b is not affected by the network infrastructure, i.e., it is the same for pico cells and femto cells. However, the optimal ζ_b increases as the load on the network increases. In fact, when more UEs associate with each SAPs, more data need to

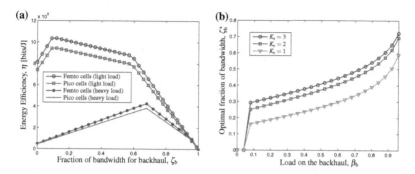

Fig. 3.5 Effect of network load on energy efficiency: **a** Energy efficiency of a small cell network that uses pico cells and femto cells, respectively, versus fraction of bandwidth ζ_b allocated to the backhaul, under different load conditions; and **b** Optimal fraction of bandwidth to be allocated to the backhaul versus load on the backhaul, for various values of the number of UEs per SAP, K_s

be forwarded from MBSs to SAPs through the wireless backhaul in order to meet the rate demand. In summary, the figure shows that irrespective of the deployment strategy, an optimal backhaul bandwidth allocation that depends on the network load can be highly beneficial to the energy efficiency of a small cell network.

In Fig. 3.5b, we plot the optimal value ζ_b^* for the fraction of bandwidth to be allocated to the backhaul as a function of the load on the backhaul β_b. We consider femto cell deployment for three different values of the number of UEs per SAP, K_s. Consistently with Fig. 3.5a, this figure shows that the optimal fraction of bandwidth ζ_b^* to be allocated to the wireless backhaul increases as β_b or K_s increase, since the load on the wireless backhaul becomes heavier and more resources are needed to meet the data rate demand.

In Fig. 3.6, the energy efficiency of the small cell network is plotted as a function of the MBS transmit power under different deployment strategies and load conditions. From the figure we can see that there is an optimal value for the transmit power, and this is given by a tradeoff between the data rate that the wireless backhaul can support and the power consumption incurred. Under spatial multiplexing, the data rate of the network is affected by the number of scheduled UEs per base station antenna, which we denote as the network load. As a consequence, the network load highly affects the data rate, and in turn affects the energy efficiency.

In Fig. 3.7, we plot the energy efficiency of the small cell network versus the number of SAPs per MBS. We consider four scenarios: (i) optimal bandwidth allocation, where the fraction of bandwidth ζ_b for the backhaul is chosen as the one that maximizes the overall energy efficiency; (ii) proportional bandwidth allocation, where the fraction of bandwidth allocated to the backhaul is equal to the fraction of load on the backhaul, i.e., $\zeta_b = \frac{K_b K_s}{K_m + K_b K_s}$ [35]; (iii) fixed bandwidth allocation, where the bandwidth is equally divided between macro- and small-cell links and wireless backhaul, i.e., $\zeta_b = 0.5$; and (iv) one-tier cellular network, where no SAPs or wireless backhaul are used at all, and all the bandwidth is allocated to the macro cell link, i.e., $\zeta_b = 0$. Figure 3.7 shows that in a two-tier small cell network there is

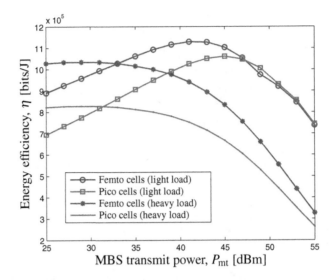

Fig. 3.6 Energy efficiency of a small cell network that uses pico cells and femto cells, respectively, versus power allocated to the backhaul, under different load conditions

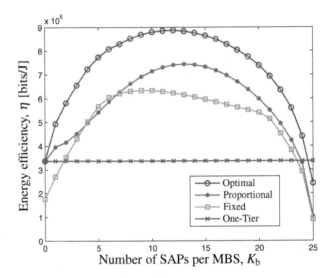

Fig. 3.7 Energy efficiency versus number of SAPs per MBS under various bandwidth allocation schemes

an optimal number of SAPs associated to each MBS via the wireless backhaul that maximizes the energy efficiency. Such number is given by a tradeoff between the data rate that the SAPs can provide to the UEs and the total power consumption. This figure also indicates that a two-tier small cell network with wireless backhaul can be less energy efficient than a single tier cellular network if the wireless backhaul is not supported well. However, when the backhaul bandwidth is optimally allocated, the small cell network can achieve a significant gain over a one-tier deployment in terms of energy efficiency.

3.6 Concluding Remarks

In this chapter, we undertook an analytical study for the energy-efficient design of small cell network with a wireless backhaul. By combining stochastic geometry and random matrix theory, we developed a framework that is general and accounts for uplink and downlink transmissions, spatial multiplexing, and resource allocation between radio access links and backhaul. The framework allows to explicitly characterize the power consumption of the small cell network due to the signal processing operations in macro cells, small cells, and wireless backhaul, as well as the data rates and ultimately the energy efficiency of the whole network.

Our results revealed that, irrespective of the deployment strategy, it is critical to control the network load in order to maintain a high energy efficiency. Moreover, a two-tier small cell network with wireless backhaul can achieve a significant energy efficiency gain over a one-tier deployment, as long as the bandwidth division between radio access links and wireless backhaul is optimally designed.

Appendix

Proof of Lemma 3.1

The channel matrix between a MBS to its K_m associated UEs can be written as $\hat{\mathbf{H}} = \mathbf{L}^{\frac{1}{2}}\mathbf{H}$, where $\mathbf{L} = \text{diag}\{L_1^{-1}, \ldots, L_{K_m}^{-1}\}$, with $L_i = r_i^\alpha / S_i$ being the path loss from the MBS to its ith UE, where r_i is the corresponding distance and S_i denotes the shadowing, $\mathbf{H} = [\mathbf{h}_1, \ldots, \mathbf{h}_{K_m}]^T$ is the $K_m \times M_m$ small-scale fading matrix, with $\mathbf{h}_i \sim CN(\mathbf{0}, \mathbf{I})$. The ZF precoder is then given by $\mathbf{W} = \xi\hat{\mathbf{H}}^*(\hat{\mathbf{H}}\hat{\mathbf{H}}^*)^{-1}$, where $\xi^2 = 1/\text{tr}[(\hat{\mathbf{H}}^*\hat{\mathbf{H}})^{-1}]$ normalizes the transmit power [56]. In the following, we use the notation Φ^U as Φ^D to denote the subsets of Φ that transmit in uplink and downlink, respectively, we further denote $\mathcal{U}_x$ as the set of UEs that are associated with access point x, and denote $\hat{x}$ as the transmitter that locates closest to the origin. Since the locations of MBSs and SAPs follow a stationary PPP, we can apply the Slivnyark's theorem [54], which implies that it is sufficient to evaluate the SINR of a typical UE at the origin. As such, by noticing that under dynamic TDD, every wireless link experiences interference from the downlink transmitting MBSs and SAPs, and from the uplink transmitting UEs, the downlink SINR between a typical UE at the origin and its serving MBS can be written as

$$\gamma_m^{DL} = \frac{P_{mt}|\mathbf{h}_{\hat{x}_m,o}^* \mathbf{w}_{\hat{x}_m,o}|^2 L_{\hat{x}_m,o}^{-1}}{I_{oc}^{mu} + I_u + \sigma^2}, \qquad (3.25)$$

where $\mathbf{h}_{\hat{x}_m,o}$ is the small-scale fading, $\mathbf{w}_{\hat{x}_m,o}$ is the ZF precoding vector, $L_{\hat{x}_m,o}$ denotes the corresponding path loss, while I_{oc}^{mu} is the aggregate interference from other cells to the MBS UE, and I_u denotes the interference from UEs, respectively given as follows:

$$I_{oc}^{mu} = \sum_{x_m \in \Phi_m^D \setminus \hat{x}_m} \frac{P_{mt} g_{x_m,o}}{K_m L_{x_m,o}} + \sum_{x_s \in \Phi_s^D} \frac{P_{st} g_{x_s,o}}{K_s L_{x_s,o}} \qquad (3.26)$$

and

$$I_u = \sum_{x_u \in \Phi_u^U} \frac{P_{ut}|h_{x_u,o}|^2}{L_{x_u,o}}, \qquad (3.27)$$

whereas $g_{x_m,o}$ and $g_{x_s,o}$ represent the effective small-scale fading from the interfering MBS x_m and SAP x_s to the origin, respectively, given by [57]

$$g_{x_m,o} = \sum_{u \in \mathcal{U}_{x_m}} K_m |\mathbf{h}_{x_m,o}^* \mathbf{w}_{x_m,u}|^2 \sim \Gamma(K_m, 1) \qquad (3.28)$$

and

$$g_{x_s,o} = \sum_{u \in \mathcal{U}_{x_s}} K_s |\mathbf{h}_{x_s,o}^* \mathbf{w}_{x_s,u}|^2 \sim \Gamma(K_s, 1). \qquad (3.29)$$

By conditioning on the interference, when $K_m, M_m \to \infty$ with $\beta_m = K_m/M_m < 1$, the SINR under ZF precoding converges to [39]

$$\gamma_m^{DL} \to \bar{\gamma}_m^{DL} = \frac{P_{mt} M_m}{(I_{oc}^{mu} + I_u + \sigma^2) \sum_{j=1}^{K_m} e_j^{-1}}, \quad a.s. \qquad (3.30)$$

where e_i is the solution of the fixed point equation

$$\frac{L_{\hat{x}_m,u_i}^{-1}}{e_i} = 1 + \frac{J}{M_m}, \quad i = 1, 2, \ldots, K_m \qquad (3.31)$$

with $J = \sum_{j=1}^{K_m} L_{\hat{x}_m,u_j}^{-1} e_j^{-1}$. By summing (3.31) over i we obtain

$$J = K_m + \frac{K_m}{M_m} J. \qquad (3.32)$$

Solving the equation above results in $J = K_m M_m/(M_m - K_m)$, and by substituting the value of J into (3.31) we can have

$$\frac{1}{\bar{e}_i} = \frac{M_{\mathrm{m}}}{M_{\mathrm{m}} - K_{\mathrm{m}}} \cdot L_{\hat{x}_{\mathrm{m}}, u_i}, \tag{3.33}$$

which substituted into (3.30) yields

$$\bar{\gamma}_{\mathrm{m}}^{\mathrm{DL}} = \frac{(1 - \beta_{\mathrm{m}}) \, M_{\mathrm{m}} P_{\mathrm{mt}}}{\left(I_{\mathrm{oc}}^{\mathrm{mu}} + I_{\mathrm{u}} + \sigma^2\right) \sum_{j=1}^{K_{\mathrm{m}}} L_{\hat{x}_{\mathrm{m}}, u_j}}. \tag{3.34}$$

Notice that $\{L_{\hat{x}_{\mathrm{m}}, u_j}\}_{j=1}^{K_{\mathrm{m}}}$ is an independent i.i.d. sequence with finite first moment, given by

$$\mathbb{E}\left[L_{\hat{x}_{\mathrm{m}}, u_j}\right] = \Gamma\left(1 + \frac{1}{\delta}\right) G_{\mathrm{m}}^{-1} < \infty,$$

by applying the strong law of large numbers (SLLN) to (3.34), we have

$$\bar{\gamma}_{\mathrm{m}}^{\mathrm{DL}} \to \frac{(1 - \beta_{\mathrm{m}}) \, G_{\mathrm{m}}^{1/\delta} P_{\mathrm{mt}}}{\beta_{\mathrm{m}} \Gamma\left(1 + \frac{1}{\delta}\right)\left(I_{\mathrm{oc}}^{\mathrm{mu}} + I_{\mathrm{u}} + \sigma^2\right)}, \quad a.s. \tag{3.35}$$

As such, using the continuous mapping theorem and the lemma in [58], we can compute the ergodic rate as

$$\begin{aligned}
\mathbb{E}\left[\log_2\left(1 + \bar{\gamma}_{\mathrm{m}}^{\mathrm{DL}}\right)\right] &= \frac{1}{\ln 2} \mathbb{E}\left[\ln\left(1 + \frac{v_{\mathrm{m}}^{\mathrm{D}}}{I_{\mathrm{oc}}^{\mathrm{mu}} + I_{\mathrm{u}} + \sigma^2}\right)\right] \\
&= \int_0^\infty \frac{e^{-\sigma^2 z}}{z \ln 2}\left(1 - e^{-v_{\mathrm{m}}^{\mathrm{D}} z}\right) \mathbb{E}\left[e^{-z I_{\mathrm{u}}}\right] \mathbb{E}\left[e^{-z I_{\mathrm{oc}}^{\mathrm{mu}}}\right] dz. \tag{3.36}
\end{aligned}$$

Due to the composition of independent PPPs and the displacement theorem [59], the interference I_{u} follows a homogeneous PPP with spatial density $\tilde{\lambda}_{\mathrm{u}} = (1 - \tau_{\mathrm{m}}) \lambda_{\mathrm{m}} K_{\mathrm{m}} + (1 - \tau_{\mathrm{s}}) \lambda_{\mathrm{s}} K_{\mathrm{s}}$, and the corresponding Laplace transform is given as [54]

$$\mathbb{E}\left[e^{-z I_{\mathrm{u}}}\right] = \exp\left(-\frac{2\pi^2 \tilde{\lambda}_{\mathrm{u}} \mathbb{E}[S_{\mathrm{D}}^{\frac{2}{\alpha}}] P_{\mathrm{ut}}^{\frac{2}{\alpha}} z^{\frac{2}{\alpha}}}{\alpha \sin\left(\frac{2\pi}{\alpha}\right)}\right). \tag{3.37}$$

As for the Laplace transform of $I_{\mathrm{oc}}^{\mathrm{mu}}$, the conditional Laplace transform on $L_{\hat{x}_{\mathrm{m}}, o}$ can be computed as

$$\begin{aligned}
&\mathbb{E}\left[e^{-z I_{\mathrm{oc}}^{\mathrm{mu}}} \,\middle|\, L_{\hat{x}_{\mathrm{m}}, o} = t\right] \\
&= \exp\left(-\tau_{\mathrm{m}} a_{\mathrm{m}} \mathscr{C}_{\alpha, K_{\mathrm{m}}}(z P_{\mathrm{mt}}, t)\left(\frac{z P_{\mathrm{mt}}}{K_{\mathrm{m}}}\right)^\delta - \tau_{\mathrm{s}} a_{\mathrm{s}} \mathscr{C}_{\alpha, K_{\mathrm{s}}}(z P_{\mathrm{mt}}, t)\left(\frac{z P_{\mathrm{st}}}{K_{\mathrm{s}}}\right)^\delta\right). \tag{3.38}
\end{aligned}$$

Notice that $L_{\hat{x}_m,o}$ has its distribution given by (3.13), and the rate R_m^{DL} given as

$$R_m^{DL} = (1 - \zeta_b)\, \mathbb{E}\left[\log_2\left(1 + \bar{\gamma}_m^{DL}\right)\right],\tag{3.39}$$

substituting (3.37) and (3.38) into (3.36), and decondition $L_{\hat{x}_m,o}$ with respect to (3.13) we have the corresponding result.

Proof of Lemma 3.2

Let us consider a UE transmitting in uplink to a typical MBS located at the origin, which employs a ZF receive filter $\mathbf{r}^*_{o,\hat{x}_u} = \hat{\mathbf{h}}^*_{o,\hat{x}_u}(\sum_{u\in\mathscr{U}_o}\hat{\mathbf{h}}_{o,u}\hat{\mathbf{h}}^*_{o,u})^{-1}$ [56], the SINR is then given by

$$\gamma_m^{UL} = \frac{P_{ut}L_{o,\hat{x}_u}^{-1}|\mathbf{r}^*_{o,\hat{x}_u}\mathbf{h}_{o,\hat{x}_u}|^2}{(I_{oc}^{mbs} + I_u + \sigma^2)\,\|\mathbf{r}_{o,\hat{x}_u}\|^2},\tag{3.40}$$

where I_{oc}^{mbs} denotes the interference from other cells received at the MBS. By conditioning on the interference, when $K_m, M_m \to \infty$ with $\beta_m = K_m/M_m < 1$, the SINR above converges to [39]

$$\gamma_m^{UL} \to \bar{\gamma}_m^{UL} = \frac{P_{ut}M_m(1 - \beta_m)L_{o,\hat{x}_u}^{-1}}{I_{oc}^{mbs} + I_u + \sigma^2},\quad a.s.\tag{3.41}$$

By using the continuous mapping theorem [58], the uplink ergodic rate can be calculated as

$$\mathbb{E}\left[\log_2\left(1 + \bar{\gamma}_m^{UL}\right)\right] = \frac{1}{\ln 2}\mathbb{E}\left[\ln\left(1 + \frac{v_m^U L_{o,x_u}^{-1}}{I_{oc}^{mbs} + I_u + \sigma^2}\right)\right]$$

$$= \int_0^\infty\int_0^\infty \frac{e^{-\sigma^2 z}}{z\ln 2}\left(1 - e^{-zv_m^U/t}\right)\mathbb{E}\left[e^{-zI_u}\right]\mathbb{E}\left[e^{-zI_{oc}^{mbs}}\right] f_{L_m}(t)dzdt.\tag{3.42}$$

The Laplace transform of I_{oc}^{mbs} can be computed as

$$\mathbb{E}\left[e^{-zI_{oc}^{mbs}}\right]$$
$$= \exp\left(-\frac{\Gamma(1+\delta)\,\delta\pi^2 z^\delta}{\sin(\delta\pi)}\left[\frac{\tau_m a_m P_{mt}^\delta \prod_{i=1}^{K_m-1}(i+\delta)}{\Gamma(K_m)K_m^\delta} + \frac{\tau_s a_s P_{st}^\delta \prod_{i=1}^{K_s-1}(i+\delta)}{\Gamma(K_s)K_s^\delta}\right]\right).\tag{3.43}$$

On the other hand, to consider the uplink interference from UEs, we use the result in [60] where the path loss from MBS UEs and SAP UEs are modeled as two independent inhomogeneous PPP with intensity measure being

$$\Lambda_{mu}^{(m)}(dx) = \delta a_m x^{\delta-1}\left[1 - \exp\left(-G_m x^{\delta}\right)\right],\tag{3.44}$$

$$\Lambda_{su}^{(m)}(dx) = \delta a_s x^{\delta-1}\left[1 - \exp\left(-G_m x^{\delta}\right)\right].\tag{3.45}$$

The Laplace transform of the UE interference can then be calculated as

$$\mathbb{E}[e^{-zI_u}] = \exp\left(-(1-\tau_m)K_m\int_0^{\infty}\frac{\Lambda_{mu}^{(m)}(dx)}{1+z^{-1}x/P_{ut}} - (1-\tau_s)K_s\int_0^{\infty}\frac{\Lambda_{su}^{(m)}(dx)}{1+z^{-1}x/P_{ut}}\right)$$

$$= \exp\left(-\tilde{\lambda}_u\pi\mathbb{E}[S_D^{\delta}]\int_0^{\infty}\frac{1-e^{-G_m u}}{1+z^{-1}u^{\frac{1}{\delta}}/P_{ut}}du\right).\tag{3.46}$$

As such, noticing that

$$R_m^{UL} = (1-\zeta_b)\,\mathbb{E}\left[\log_2\left(1+\bar{\gamma}_m^{UL}\right)\right]\tag{3.47}$$

the result follows by substituting (3.43) and (3.46) into (3.42).

Proof of Lemma 3.7

The average rate for a typical UE located at the origin is given by

$$R = A_m R_m + A_s R_s,\tag{3.48}$$

where R_m and R_s are the data rates when the UE associates to a MBS and a SAP, respectively, given by

$$R_m = \tau_m R_m^{DL} + (1-\tau_m)R_m^{UL}\tag{3.49}$$

and

$$R_s = \tau_s \min\left\{R_s^{DL}, R_b^{DL}\right\} + (1-\tau_s)\min\left\{R_s^{UL}, R_b^{UL}\right\}.\tag{3.50}$$

As each MBS and each SAP serve K_m and K_s UEs, respectively, the total density of active UEs is given by $K_m\lambda_m + K_s\lambda_s$. Let B be the available bandwidth, the sum rate per area is obtained as $\mathscr{R} = (K_m\lambda_m + K_s\lambda_s)\,BR$. Lemma 3.7 then follows from Lemmas 3.1 to 3.6 and by the continuous mapping theorem.

References

1. G. Auer, V. Giannini, C. Desset, I. Godor, P. Skillermark, M. Olsson, M. Imran, D. Sabella, M. Gonzalez, O. Blume, and A. Fehske, "How much energy is needed to run a wireless network?" *IEEE Wireless Commun.*, vol. 18, no. 5, pp. 40–49, Oct. 2011.
2. Y. Chen, S. Zhang, S. Xu, and G. Y. Li, "Fundamental trade-offs on green wireless networks," *IEEE Commun. Mag.*, vol. 49, no. 6, pp. 30–37, Jun. 2011.
3. G. Y. Li, Z. Xu, C. Xiong, C. Yang, S. Zhang, Y. Chen, and S. Xu, "Energy-efficient wireless communications: Tutorial, survey, and open issues," *IEEE Trans. Wireless Commun.*, vol. 18, no. 6, pp. 28–35, Dec. 2011.
4. D. Feng, C. Jiang, G. Lim, L. J. Cimini Jr, G. Feng, and G. Y. Li, "A survey of energy-efficient wireless communications," *IEEE Commun. Surveys and Tutorials*, vol. 15, no. 1, pp. 167–178, Feb. 2013.
5. R. Hu and Y. Qian, "An energy efficient and spectrum efficient wireless heterogeneous network framework for 5G systems," *IEEE Commun. Mag.*, vol. 52, no. 5, pp. 94–101, May 2014.
6. G. Geraci, M. Wildemeersch, and T. Q. S. Quek, "Energy efficiency of distributed signal processing in wireless networks: A cross-layer analysis," *IEEE Trans. Signal Process.*, vol. 64, no. 4, pp. 1034–1047, Feb. 2016.
7. H. H. Yang, J. Lee, and T. Q. S. Quek, "Heterogeneous cellular network with energy harvesting-based D2D communication," *IEEE Trans. Wireless Commun.*, vol. 15, no. 2, pp. 1406–1419, Feb. 2016.
8. T. Q. S. Quek, G. de la Roche, I. Güvenç, and M. Kountouris, *Small cell networks: Deployment, PHY techniques, and resource management*. Cambridge University Press, 2013.
9. J. G. Andrews, H. Claussen, M. Dohler, S. Rangan, and M. C. Reed, "Femtocells: Past, present, and future," *IEEE J. Sel. Areas Commun.*, vol. 30, no. 3, pp. 497–508, Apr. 2012.
10. J. Hoydis, M. Kobayashi, and M. Debbah, "Green small-cell networks," *IEEE Vehicular Technology Mag.*, vol. 6, no. 1, pp. 37–43, Mar. 2011.
11. Q. Ye, B. Rong, Y. Chen, M. Al-Shalash, C. Caramanis, and J. G. Andrews, "User association for load balancing in heterogeneous cellular networks," *IEEE Trans. Wireless Commun.*, vol. 12, no. 6, pp. 2706–2716, Jun. 2013.
12. H. S. Dhillon, M. Kountouris, and J. G. Andrews, "Downlink MIMO hetnets: Modeling, ordering results and performance analysis," *IEEE Trans. Wireless Commun.*, vol. 12, no. 10, pp. 5208–5222, Oct. 2013.
13. Small Cell Forum, "Backhaul technologies for small cells," white paper, document 049.05.02, Feb. 2014.
14. H. S. Dhillon and G. Caire, "Information theoretic upper bound on the capacity of wireless backhaul networks," in *Proc. IEEE Int. Symp. on Inform. Theory*, Honolulu, HI, Jun. 2014, pp. 251–255.
15. H. S. Dhillon and G. Caire, "Scalability of line-of-sight massive MIMO mesh networks for wireless backhaul," in *Proc. IEEE Int. Symp. on Inform. Theory*, Honolulu, HI, Jun. 2014, pp. 2709–2713.
16. L. Sanguinetti, A. L. Moustakas, and M. Debbah, "Interference management in 5G reverse TDD HetNets: A large system analysis," *IEEE J. Sel. Areas Commun.*, vol. 33, no. 6, pp. 1–1, Mar. 2015.
17. J. Andrews, "Seven ways that HetNets are a cellular paradigm shift," *IEEE Commun. Mag.*, vol. 51, no. 3, pp. 136–144, Mar. 2013.
18. H. Claussen, "Future cellular networks," Alcatel-Lucent, Apr. 2012.
19. A. J. Fehske, F. Richter, and G. P. Fettweis, "Energy efficiency improvements through micro sites in cellular mobile radio networks," in *Proc. IEEE Global Telecomm. Conf. Workshops*, Honolulu, HI, Dec. 2009, pp. 1–5.

20. C. Li, J. Zhang, and K. Letaief, "Throughput and energy efficiency analysis of small cell networks with multi-antenna base stations," *IEEE Trans. Wireless Commun.*, vol. 13, no. 5, pp. 2505–2517, May 2014.

21. M. Wildemeersch, T. Q. S. Quek, C. H. Slump, and A. Rabbachin, "Cognitive small cell networks: Energy efficiency and trade-offs," *IEEE Trans. Commun.*, vol. 61, no. 9, pp. 4016–4029, Sep. 2013.

22. Y. S. Soh, T. Q. S. Quek, M. Kountouris, and H. Shin, "Energy efficient heterogeneous cellular networks," *IEEE J. Sel. Areas Commun.*, vol. 31, no. 5, pp. 840–850, Apr. 2013.

23. S. Navaratnarajah, A. Saeed, M. Dianati, and M. A. Imran, "Energy efficiency in heterogeneous wireless access networks," *IEEE Wireless Commun.*, vol. 20, no. 5, pp. 37–43, Oct. 2013.

24. E. Björnson, M. Kountouris, and M. Debbah, "Massive MIMO and small cells: Improving energy efficiency by optimal soft-cell coordination," in *Proc. IEEE Int. Conf. on Telecommun.*, Casablanca, Morocco, May 2013, pp. 1–5.

25. X. Ge, H. Cheng, M. Guizani, and T. Han, "5G wireless backhaul networks: Challenges and research advances," *IEEE Network*, vol. 28, no. 6, pp. 6–11, Dec. 2014.

26. S. Tombaz, P. Monti, F. Farias, M. Fiorani, L. Wosinska, and J. Zander, "Is backhaul becoming a bottleneck for green wireless access networks?" in *Proc. IEEE Int. Conf. Commun.*, Sydney, Australia, Jun. 2014, pp. 4029–4035.

27. S. Tombaz, P. Monti, K. Wang, A. Vastberg, M. Forzati, and J. Zander, "Impact of backhauling power consumption on the deployment of heterogeneous mobile networks," in *Proc. IEEE Global Telecomm. Conf.*, Houston, TX, Dec. 2011, pp. 1–5.

28. M. Sanchez-Fernandez, S. Zazo, and R. Valenzuela, "Performance comparison between beam-forming and spatial multiplexing for the downlink in wireless cellular systems," *IEEE Trans. Wireless Commun.*, vol. 6, no. 7, pp. 2427–2431, Jul. 2007.

29. H. Shirani-Mehr, G. Caire, and M. J. Neely, "MIMO downlink scheduling with non-perfect channel state knowledge," *IEEE Trans. Commun.*, vol. 58, no. 7, pp. 2055–2066, Jul. 2010.

30. V. Chandrasekhar and J. G. Andrews, "Spectrum allocation in tiered cellular networks," *IEEE Trans. Commun.*, vol. 57, no. 10, pp. 3059–3068, Oct. 2009.

31. W. C. Cheung, T. Q. S. Quek, and M. Kountouris, "Throughput optimization, spectrum alloca-tion, and access control in two-tier femtocell networks," *IEEE J. Sel. Areas Commun.*, vol. 30, no. 3, pp. 561–574, Apr. 2012.

32. T. Zahir, K. Arshad, A. Nakata, and K. Moessner, "Interference management in femtocells," *IEEE Commun. Surveys and Tutorials*, vol. 15, no. 1, pp. 293–311, 2013.

33. M. Peng, C. Wang, J. Li, H. Xiang, and V. Lau, "Recent advances in underlay heterogeneous networks: Interference control, resource allocation, and self-organization," *IEEE Commun. Surveys and Tutorials*, vol. 17, no. 2, pp. 700–729, May 2015.

34. H. S. Dhillon and G. Caire, "Wireless backhaul networks: Capacity bound, scalability analysis and design guidelines," *IEEE Trans. Wireless Commun.*, vol. 14, no. 11, pp. 6043–6056, Nov. 2015.

35. S. Singh, M. N. Kulkarni, A. Ghosh, and J. G. Andrews, "Tractable model for rate in self-backhauled millimeter wave cellular networks," *IEEE J. Sel. Areas Commun.*, vol. 33, no. 10, pp. 2196–2211, Oct. 2015.

36. Z. Shen, A. Khoryaev, E. Eriksson, and X. Pan, "Dynamic uplink-downlink configuration and interference management in TD-LTE," *IEEE Commun. Mag.*, vol. 50, no. 11, pp. 51–59, Nov. 2012.

37. M. Ding, D. Lopez Perez, A. V. Vasilakos, and W. Chen, "Dynamic TDD transmissions in homogeneous small cell networks," in *Proc. IEEE Int. Conf. Commun.*, Sydney, Australia, Jun. 2014, pp. 616–621.

38. Q. H. Spencer, C. B. Peel, A. L. Swindlehurst, and M. Haardt, "An introduction to the multi-user MIMO downlink," *IEEE Commun. Mag.*, vol. 42, no. 10, pp. 60–67, Oct. 2004.

39. S. Wagner, R. Couillet, M. Debbah, and D. T. Slock, "Large system analysis of linear precoding in correlated miso broadcast channels under limited feedback," *IEEE Trans. Inf. Theory*, vol. 58, no. 7, pp. 4509–4537, Jul. 2012.

40. G. Geraci, M. Egan, J. Yuan, A. Razi, and I. B. Collings, "Secrecy sum-rates for multi-user MIMO regularized channel inversion precoding," *IEEE Trans. Commun.*, vol. 60, no. 11, pp. 3472–3482, Nov. 2012.

41. G. Geraci, A. Y. Al-Nahari, J. Yuan, and I. B. Collings, "Linear precoding for broadcast channels with confidential messages under transmit-side channel correlation," *IEEE Commun. Lett.*, vol. 17, no. 6, pp. 1164–1167, Jun. 2013.

42. G. Geraci, R. Couillet, J. Yuan, M. Debbah, and I. B. Collings, "Large system analysis of linear precoding in MISO broadcast channels with confidential messages," *IEEE J. Sel. Areas Commun.*, vol. 31, no. 9, pp. 1660–1671, Sep. 2013.

43. E. Björnson, L. Sanguinetti, J. Hoydis, and M. Debbah, "Optimal design of energy-efficient multi-user MIMO systems: Is massive MIMO the answer?" *IEEE Trans. Wireless Commun.*, vol. 14, no. 6, pp. 3059 – 3075, Jun. 2015.

44. A. Mezghani and J. A. Nossek, "Power efficiency in communication systems from a circuit perspective," in *IEEE Int. Symp. on Circuits and Systems*, Rio de Janeiro, May 2011, pp. 1896–1899.

45. S. Tombaz, A. Västberg, and J. Zander, "Energy-and cost-efficient ultra-high-capacity wireless access," *IEEE Wireless Commun.*, vol. 18, no. 5, pp. 18–24, Oct. 2011.

46. H. Yang and T. L. Marzetta, "Total energy efficiency of cellular large scale antenna system multiple access mobile networks," in *IEEE Online GreenCom*, Oct. 2013, pp. 27–32.

47. T. Chen, H. Kim, and Y. Yang, "Energy efficiency metrics for green wireless communications," in *Proc. IEEE Int. Conf. Wireless Commun. and Signal Processing*, Suzhou, China, 2010, pp. 1–6.

48. S. Singh, H. S. Dhillon, and J. G. Andrews, "Offloading in heterogeneous networks: Modeling, analysis, and design insights," *IEEE Trans. Wireless Commun.*, vol. 12, no. 5, pp. 2484–2497, May 2013.

49. G. Caire and S. Shamai, "On the achievable throughput of a multiantenna gaussian broadcast channel," *IEEE Trans. Inf. Theory*, vol. 49, no. 7, pp. 1691–1706, Jul. 2003.

50. Q. H. Spencer, A. L. Swindlehurst, and M. Haardt, "Zero-forcing methods for downlink spatial multiplexing in multiuser MIMO channels," *IEEE Trans. Signal Process.*, vol. 52, no. 2, pp. 461–471, Feb. 2004.

51. T. Yoo and A. Goldsmith, "On the optimality of multiantenna broadcast scheduling using zero-forcing beamforming," *IEEE J. Sel. Areas Commun.*, vol. 24, no. 3, pp. 528–541, Mar. 2006.

52. G. Auer, V. Giannini, C. Desset, I. Godor, P. Skillermark, M. Olsson, M. A. Imran, D. Sabella, M. J. Gonzalez, O. Blume *et al.*, "How much energy is needed to run a wireless network?" *IEEE Wireless Commun. Mag.*, vol. 18, no. 5, pp. 40–49, Oct. 2011.

53. B. Debaillie, C. Desset, and F. Louagie, "A flexible and future-proof power model for cellular base stations," in *IEEE Vehicular Technology Conference*, Glasgow, Scotland, May 2015, pp. 1–7.

54. F. Baccelli and B. Blaszczyszyn, *Stochastic Geometry and Wireless Networks. Volumn I: Theory.*Now Publishers, 2009.

55. H. H. Yang, G. Geraci, and T. Q. S. Quek, "Rate analysis of spatial multiplexing in MIMO heterogeneous networks with wireless backhaul," in *Proc. IEEE Int. Conf. Acoustics, Speech, and Signal Processing*, Shanghai, China, Mar. 2016.

56. D. N. C. Tse and P. Viswanath, *Fundamentals of Wireless Communication*. Cambridge University Press, 2005.

57. G. Geraci, H. S. Dhillon, J. G. Andrews, J. Yuan, and I. Collings, "Physical layer security in downlink multi-antenna cellular networks," *IEEE Trans. Commun.*, vol. 62, no. 6, pp. 2006–2021, Jun. 2014.

58. K. A. Hamdi, "A useful lemma for capacity analysis of fading interference channels," *IEEE Trans. Commun.*, vol. 58, no. 2, pp. 411–416, Feb. 2010.
59. M. Haenggi, *Stochastic geometry for wireless networks*. Cambridge University Press, 2012.
60. S. Singh, X. Zhang, and J. Andrews, "Joint rate and SINR coverage analysis for decoupled uplink-downlink biased cell associations in hetnets," *IEEE Trans. Wireless Commun.*, vol. 14, no. 10, pp. 5360–5373, Oct. 2015.

Chapter 4
Conclusions

Abstract In this chapter, we summarize our contribution in this book and provide several new directions for the research of future work.

The concept of massive MIMO enables the communication system to be scaled up into a regime where antenna arrays at base stations greatly exceed the number of active end users, thus resulting in tremendous diversity gain. Such gain not only provides opportunity to improve capacity through spatial multiplexing, but also powers new directions to rethink the network design.

This book provides an overview for new aspects of system design that utilizes the large amount of spatial dimensions in massive MIMO to enhance network performance. Particularly, we propose two approaches to exploit the vast antenna array: (1) by exploiting spatial degree of freedom to suppress interference from BSs to the most vulnerable cell-edge UEs, we propose the cell-edge aware zero forcing (CEA-ZF) precoding scheme; (2) by noticing the tremendous diversity gain at the antenna array, we propose applying massive MIMO for wireless backhaul in a two-tier small cell network such that one macro base station can transmit to several small access points simultaneously through spatial multiplexing. In order to take a complete treatment of all the key features of a wireless network, such as wireless channel, random base stations location, and large antenna array, we combine tools from random matrix theory and stochastic geometry to develop an analytical frameworks which is general and accounts for all the key network parameters. With the analytical results, we show that CEA-ZF outperforms the conventional CEU-ZF in terms of aggregated per-cell data rate, coverage probability, and a significant 95 %-likely, or edge-user rate. On the other hand, we show that in a small cell network with wireless backhaul, irrespective of the deployment strategy, it is critical to control the network load in order to maintain a high energy efficiency. Moreover, a two-tier small cell network with wireless backhaul can achieve a significant energy efficiency gain over a one-tier deployment, as long as the bandwidth division between radio access links and wireless backhaul is optimally allocated.

In the future work, we can extend the framework developed in this book to the investigation of pilot contamination issues in massive MIMO cellular network, also the analysis with full duplex in a small cell network.

© The Author(s) 2017   61
H.H. Yang and T.Q.S. Quek, *Massive MIMO Meets Small Cell*, SpringerBriefs
in Electrical and Computer Engineering, DOI 10.1007/978-3-319-43715-6_4

Printed in the United States
By Bookmasters